AF314006

The OFFICIAL Guide
ETS
Educational Testing Service
We prepare the tests—let us help prepare you!
THE PRAXIS SERIES
Professional Assessments for Beginning Teachers
A Guide to the NTE
Mathematics
Specialty Area Test

NATURE AND PURPOSE OF NTE PROGRAMS

Current NTE tests aim to measure in an objective and standardized manner academic knowledge and skills important for beginning teachers and other education professionals. Information about academic knowledge and skills of the type gained from NTE tests is significant but limited. It is significant because knowledge and skills gained through academic preparation for a profession are obviously related to the future professional's performance. The information is limited because the academic knowledge and skills tested do not include many other elements important to professional performance, such as dedication and motivation, human relations skills, perseverance, caring, and those teaching skills learned only through practice in classroom settings. Thus, at best, NTE tests can properly be said to measure only a part, but a necessary part, of preparation for a profession.

The Pre-Professional Skills Tests (PPST), the Core Battery, and the Specialty Area tests were developed based on academic knowledge and skills important for beginning teachers and have been used to determine that individuals have satisfied one of several criteria necessary to make them eligible for initial certification. (Review has shown that the Pre-Professional Skills Tests and the Communication Skills test cover academic knowledge and skills that are typically taught before or during the first two years of college and that the Professional Knowledge test covers material typically taught in teacher education programs.) In recent years, states have adopted NTE tests as part of career ladder and master teacher plans and as options for course equivalencies in recertification plans.

NTE Programs include:

- the Pre-Professional Skills Tests (PPST), which measure basic proficiency in reading, writing, and mathematics;
- the Core Battery, which covers areas common to most teacher education programs;
- the Specialty Area tests, which are designed to assess preparation in specific subject fields; and
- other teacher tests developed for institutions or agencies under special contract.

The Specialty Area tests measure knowledge of specific academic subjects or fields. Although most of the tests are designed to evaluate the knowledge and skills prospective teachers acquire in their academic preparation, some — such as Speech-Language Pathology and Audiology — are designed to measure examinees' preparation for careers other than teaching. Most of the Specialty Area tests are intended for college seniors; a few, however, are designed for those who have had graduate training and/or other educational experience.

This book was prepared and produced by Educational Testing Service (ETS), which develops and administers NTE Programs tests.

Although ETS conducts the NTE Programs, it is assisted and advised by professional educators from all sections of the country. The tests themselves are developed and revised periodically by ETS with the assistance of committees of recognized authorities in specific subject fields. These committees are usually appointed from nominations made by appropriate national professional associations.

The Teacher Programs Council advises and assists ETS in ongoing examination, analysis, and discussion of policy issues concerning the NTE Programs tests and other ETS programs and services for teachers. The 20-member Council is comprised of classroom teachers, school district superintendents, college and university deans or faculty, chief state school officers, and knowledgeable members of the public.

TABLE OF CONTENTS

INTRODUCTION

The purpose of this guide is to help you prepare for the NTE Mathematics Specialty Area test. The guide cannot teach you the content of the test or assure you a high score; the knowledge gained in your years of schooling is the most important factor in determining how well you will do on the test. What the book *can* do is acquaint you with question formats, help you pinpoint strong and weak areas, and assist you in honing your test-taking skills. As a result, you can effectively prepare for an actual NTE administration.

The guide includes four parts.

Part One provides an overview of the NTE Specialty Area tests and information about how they are developed. Various question formats are identified and explained. Part One also includes a section on test-taking strategies designed to help you reduce the anxiety often associated with taking a standardized test.

Part Two includes content specifications for the Mathematics test and provides sample questions with explanations of the best answers.

Part Three consists of a complete Mathematics practice test and a copy of the NTE Specialty Area answer sheet. The practice test is a disclosed (previously administered) form of the Mathematics test.

Part Four contains the answer key for the practice test and explanations of the answers. It also includes instructions for scoring the Mathematics test and computing your score.

PART ONE: AN OVERVIEW OF THE NTE SPECIALTY AREA TESTS

There are currently more than 60 NTE Specialty Area tests designed to assess knowledge and skills across a wide spectrum of disciplines. Most of the tests measure the academic competence of prospective teachers in the areas in which they plan/intend to teach. However, some tests — such as Speech-Language Pathology, Audiology, School Guidance and Counseling, and School Psychologist — measure preparation for careers other than classroom teaching.

Designed for prospective secondary school teachers, the Mathematics test was administered to more than 7,000 test takers in 1989-90.

It is two hours long and contains 120 questions. Currently, all Specialty Area tests have a multiple-choice format with five or (in the case of tests in the language area) four answer choices (options). Some tests include audiotaped sections requiring test takers to select answers in response to taped stimuli. The Mathematics test does *not* contain an audiotaped section.

How the NTE Programs Specialty Area Tests Are Developed

Before a Specialty Area test is developed, educators and other professionals determine the need for such a test. Once that determination is made, test developers must ask some critical questions:

- Who will take the test and for what purposes?
- How much should the test takers be expected to know about the field?
- How should the test takers be able to use their knowledge?
- What kinds of questions should be used, and how many of each kind?
- How long should the test be?
- What should the difficulty level of the test be?

Answers to these questions come from a committee of examiners made up of subject matter experts from across the country who meet to discuss and develop preliminary specifications for the test.

Additional subject matter experts in the field review the preliminary specifications to make sure they are adequate and comprehensive. The specifications are made final only after the committee of examiners has carefully considered the information obtained from the reviews.

The questions in an NTE Specialty Area test are prepared by ETS staff and outside experts. Every question is carefully reviewed by the ETS test development staff and editors to make sure it conforms to ETS standards. The test is then assembled by the ETS test development staff according to content and difficulty specifications established by the committee. ETS staff members, the committee members, and the subject matter experts review the assembled test to check for accuracy and appropriateness.

Before any test is administered, it must undergo a sensitivity review. This is a check to ensure that the questions truly reflect the diverse cultural nature of our society. No test will be administered if it contains questions that could be regarded as biased toward or against any subgroup of the population. The sensitivity review is conducted by ETS staff persons specially trained to identify and eliminate material that might be considered unfair or offensive to any group.

A rigorous statistical analysis is performed on each question to assure accuracy and fairness. Feedback from test takers and test center supervisors is also an important part of the ongoing review. Reports from test takers of questions thought to be ambiguously worded, for example, are thoroughly investigated. If a problem is found with a question, it is removed from the test before it is scored.

Ten Test-Taking Strategies

1. Take a watch to help you pace yourself through the test. Do not stay on any one question too long at the expense of others. Each question counts the same.

2. Familiarize yourself with the exam format and the question types. Getting acquainted with the instructions and taking the time to work through the questions in Part Three of this book should help you make the best use of your time at the actual exam.

3. Take care to read each question. Make sure you know EXACTLY what is being asked, and be sure to read all the answer choices before you choose one.

4. Remember that the general instructions ask you to choose the BEST or most appropriate answer among the choices you are given.

5. Pay close attention to the question format. The format may change in a given question. For example, some questions may stress words like NOT, LEAST, or EXCEPT. Carefully consider this type of question. It goes counter to the thinking involved in the BEST or MOST APPROPRIATE answer type. With NOT, LEAST, or EXCEPT, you are looking for the *exception*, or the choice that is incompatible with the other four.

EXAMPLE

Which of the following would NOT be a primary source for the investigation of a natural disaster?

(A) An article by a newspaper reporter who was present
(B) The memoirs of a witness
(C) A textbook account of the event
(D) A tape recording of the event
(E) A report by a police officer who was at the scene

Of the choices given, the only one that is NOT a primary source is C. So, the best answer choice is C.

A second type of special format is the "Roman numeral" format. Here is a sample question that illustrates this format:

EXAMPLE

Allies of the United States during the Second World War included which of the following countries?

   I.   **Germany**
  II. **Great Britain**
 III. **Japan**

(A) I only
(B) II only
(C) I and III only
(D) II and III only
(E) I, II, and III

In this type of format, you must choose the correct combination of options, and sometimes the options must be in a particular order or sequence. For this example, Roman numeral II, Great Britain, is the only correct answer. So, the best answer choice is B, II only.

6. Answer all the questions you are sure of first. Then, if time permits, go back to tackle the more difficult questions. You may be surprised to find that in coming back and reading a question again it will be easier to understand.

7. Go ahead and guess. The Mathematics test is scored "rights only." That means that only questions answered correctly will count toward your score. Questions answered incorrectly or omitted will not be counted. It is therefore to your advantage to answer every question, even if you guess.

8. Don't despair if you find that you do not know the answer to every question. You do not have to answer every question correctly to score well.

9. Take care in marking your answer sheet. Mark every answer clearly and boldly. Always make sure the question number on the answer sheet is the same as the question number in the test book. Erase all unintended marks completely. If the scoring machine scans two answers for a question, that question will be considered to have been omitted and will not count toward your score.

10. Never assume that any particular pattern of answer choices will or will not appear in the test (e.g., that choice A will appear most frequently or that the same answer choice cannot be correct three times in a row). The order and distribution of correct answer choices varies from test to test.

PART TWO: TEST CONTENT SPECIFICATIONS

The Mathematics test specifications were set by a Mathematics advisory committee consisting of secondary school teachers and teacher educators. The committee members were carefully selected to be representative of the various constituencies that use the test as well as the different geographical regions of the country. The committee designed the test to measure the mathematical knowledge and abilities expected of candidates who plan to teach mathematics at the secondary level.

The content specifications consist of the nine major content categories listed below. A list of some of the topics included in each category is provided along with the approximate number of questions to be allocated to each category.

I. Number Concepts and Elementary Number Theory (15-18 questions)

Topics include measurement in both the English and metric systems, estimation, factors and multiples, ratio and proportion, percent, and scientific notation.

II. Elementary and Intermediate Algebra (17-21 questions)

Topics include the fundamental operations with monomials and polynomials, algebraic fractions, solution of equations and inequalities, absolute value, quadratic and higher degree equations, systems of equations and inequalities, positive and negative integer roots and exponents, fractional exponents, complex numbers, translation, and arithmetic and geometric progressions.

III. Geometry: Informal, Euclidean, Solid, Coordinate, Transformational (20-22 questions)

Includes topics in high school geometry; visualization in 2-space and 3-space; reflections, rotations, and translations; and graphing, including lines and planes that are parallel or perpendicular to a given line or plane.

IV. Functions and Their Graphs: Algebraic, Trigonometric, Logarithmic, and Exponential (15-18 questions)

Topics include the equations and graphs of the conic sections; absolute value function; one-to-one mappings; graphs of trigonometric functions and recognition and application of trigonometric identities; domain and range of functions; odd and even functions; composite and inverse functions; and recursive definition of functions.

V. Probability and Statistics (6-8 questions)

Topics include permutations and combinations, finite and continuous probability, conditional probability, the measures of central tendency (mean, median, and mode), range, standard deviation, the normal distribution, and other simple distributions.

VI. Algebraic Topics: Structure and Systems, Linear Algebra, Abstract Algebra (7-10 questions)

Topics include the properties of groups, rings, and fields; matrices and determinants; vectors and vector spaces; and linear transformations.

VII. Calculus,* Properties of Real Numbers (8-10 questions)

Topics include limits, maxima and minima, least upper bound property, points of inflection, asymptotes, polar coordinates, evaluation and application of derivatives and integrals, continuity, mappings into or onto a set, and convergence of series.

VIII. Computer Science and Discrete Mathematics (5-8 questions)

Topics include symbolic logic, simple computer programs, congruence for integers, and more advanced topics in number theory.

IX. Professional Knowledge and the History of Mathematics (8-10 questions)

Topics include important trends in mathematics education and knowledge of professional journals; material that should have been learned in a methods course, such as methods of teaching, analyzing student errors, and curriculum; and important developments in the history of mathematics.

In addition to the nine content categories, the specifications include the following four mental-process or skill categories:

1. Knowledge and Skills (25-30 questions)
2. Understanding of Concepts (35-45 questions)
3. Proof, Application, and Problem Solving (35-45 questions)
4. Professional Understanding and Pedagogy (8-15 questions)

Reviewers often disagree about the mental-process classifications for specific questions. For example, a question that is regarded as a routine skill (category 1) type by someone who has used the skill frequently may be regarded as a category 2 type by someone who must rethink the concepts involved in order to answer the question. In view of the nebulous and partially hierarchical nature of the mental-process categories, the specifications allow for greater ranges in the numbers of questions for each of these categories. Each question in the test is classified in exactly one content category and one mental-process category. If a question tests knowledge in more than one content category, it is classified in the major category or the category that involves the most advanced content.

*Includes only single variable calculus

Sample Questions

Following are several examples of the types of questions included in the Mathematics test. A brief explanation of how the best answer was derived follows each question.

QUESTION 1

If $0 < \theta < \dfrac{\pi}{2}$, then $\sin \theta + \cos\left(\dfrac{\pi}{2} - \theta\right) =$

(A) 0
(B) 2
(C) $2 \sin \theta$
(D) $\sin 2\theta$
(E) $\sin^2 \theta$

Explanation

Since $\cos\left(\dfrac{\pi}{2} - \theta\right)$ is equal to $\sin \theta$, $\sin \theta + \cos\left(\dfrac{\pi}{2} - \theta\right) =$ $\sin \theta + \sin \theta = 2 \sin \theta$. The correct answer is C.

QUESTION 2

Each of five students made one of the following errors on a quiz. Which error suggests that the student did not know that, for all real numbers x, $\sqrt{x^2}$ is defined to be $|x|$?

(A) $\sqrt{16} = \pm 4$

(B) $\sqrt{-16} = 4$

(C) $\sqrt{x^2 + y^2} = |x + y|$

(D) $|x + y| = |x| + |y|$

(E) If $\sqrt{x^2} - \dfrac{x+1}{2} = 3$, then $2\sqrt{x^2} - x + 1 = 6$

Explanation

The student who made the error in choice A probably did not know the definition, since the statement that $\sqrt{16}$ is equal to both +4 and −4 is contrary to the definition given. With respect to choice B, since $\sqrt{-16}$ is not a real number, the definition is not applicable. With respect to choice C, the student most likely made the error because she or he thought the square root of a sum was equal to the sum of the square roots, not because she or he did not know that the radical denotes the absolute value. Choice D is irrelevant, since it has nothing to do with $\sqrt{x^2}$. The error in choice E most likely occurred because the student was careless in distributing the −1 over the sum $(x + 1)$. Clearly, the best answer is A. For questions of this type, be sure to examine all the options carefully before choosing the best answer.

QUESTION 3

A and B are two independent events. The probability that A occurs is $\dfrac{2}{5}$; the probability that B does _not_ occur is $\dfrac{5}{6}$. What is the probability that both A and B occur?

(A) $\dfrac{1}{15}$

(B) $\dfrac{2}{15}$

(C) $\dfrac{1}{3}$

(D) $\dfrac{17}{30}$

(E) $\dfrac{13}{15}$

Explanation

The probability that two independent events both occur is equal to the product of their respective proabilities of occurring. The probability that A occurs is $\dfrac{2}{5}$, and the probability that B occurs is $\left(1 - \dfrac{5}{6}\right) = \dfrac{1}{6}$. Thus, the probability that both events occur is $\dfrac{2}{5}\left(\dfrac{1}{6}\right) = \dfrac{1}{15}$, and the correct answer is A.

QUESTION 4

$x^2 \geq x^3$ if and only if

(A) $x \leq -1$
(B) $x \leq 0$
(C) $0 \leq x \leq 1$
(D) $x \leq 1$
(E) $x \geq 1$

Explanation

If $x = 0$, then $x^2 = x^3$. If $x \neq 0$, then $x^2 > 0$ and in this case the inequality can be simplified by dividing each side of the inequality by x^2, so $x \leq 1$. It follows that $x \leq 1$ for all real values of x. Thus, the best answer is D.

QUESTION 5

A square right (regular) pyramid has how many planes of symmetry?

(A) None
(B) One
(C) Two
(D) Three
(E) Four

Explanation

For a plane to be a plane of symmetry of a square right (regular) pyramid, the plane must (1) intersect the vertex of the pyramid and (2) intersect the square base of the pyramid in one of its lines of symmetry. The square base has four lines of symmetry consisting of the diagonals and the two lines that bisect the opposite sides of the square. For each line of symmetry of the square, there is one, and only one, plane containing the line and passing through the vertex of the pyramid. Thus, the number of planes of symmetry of the pyramid is equal to the number of lines of symmetry of its square base, and the correct answer is E.

QUESTION 6

If $y = \left(2x^2 + 1\right)^5$, then $\dfrac{dy}{dx} =$

(A) $5(2x^2 + 1)^4$
(B) $5x(2x^2 + 1)^4$
(C) $20x^4$
(D) $20(2x^2 + 1)^4$
(E) $20x(2x^2 + 1)^4$

Explanation

If $y = (2x^2 + 1)^5$, then, using the chain rule, $\dfrac{dy}{dx} =$
$5(2x^2 + 1)^4(4x) = 20x(2x^2 + 1)^4$, and the correct answer is E.

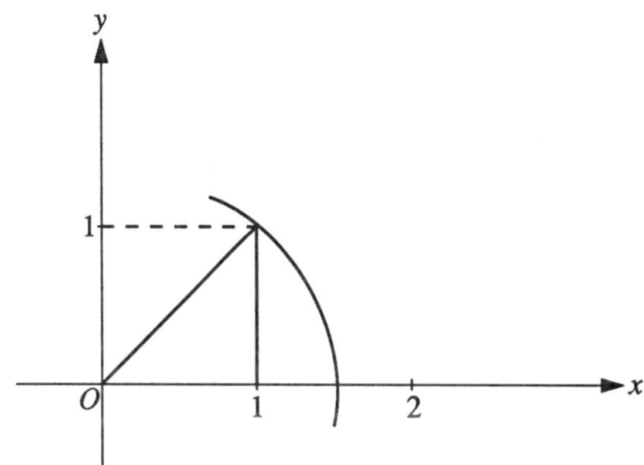

QUESTION 7

For which of the following purposes would a teacher reasonably use the sketch above?

  I. To show the location of an irrational number
 II. To illustrate the density of the irrational numbers
III. To prove the Pythagorean theorem

(A) I only
(B) II only
(C) III only
(D) I and II
(E) I and III

Explanation

With respect to I, since the diagonal of the unit square is $\sqrt{2}$ and $\sqrt{2}$ is an irrational number, the teacher could use the sketch to show the position of $\sqrt{2}$ on the x-axis. The figure does not illustrate the density of the irrational numbers, nor does it constitute a proof of the Pythagorean theorem. The best answer is A.

PART THREE: PRACTICE TEST

Now that you have been introduced to the content of the test, have considered some test-taking strategies, and have read through sample questions, you should try taking the practice test. You will probably find it helpful to simulate actual test conditions, following the directions exactly and giving yourself two hours to work on the questions. An answer sheet can be found at the end of the test. When you have finished the test, you can turn to the instructions for scoring your test in Part Four.

You will find it useful to read the explanations of the best answer choices for the questions in the practice test. These explanations, which were written by members of the committee of examiners or by ETS test developers, cover many of the issues raised by the questions they concern. Moreover, they present a fair sample of the kinds of questions asked and issues raised in the Mathematics Specialty Area test.

Keep in mind that the test you take at an actual administration may be slightly more difficult or slightly easier than the test included in this book. The content of test questions may be different, and you yourself may not perform at exactly the same level every time you take a test. Therefore, you should not expect to get the same score when you take a test at an actual administration that you get when you take the practice test.

Specialty Area Test

NTE® Programs

MATHEMATICS

TEST CODE

06017

Read the directions on the back cover.

*Do not break the seal
until you are told to do so.*

*Hand in this test book and the
answer sheet **separately** at the
conclusion of the test.*

THIS TEST BOOK MUST NOT BE TAKEN FROM THE ROOM.

MATHEMATICS

Time—120 minutes

120 Questions

Directions: Each of the questions or incomplete statements below is followed by five suggested answers or completions. Select the one that is best in each case and then fill in the corresponding lettered space on the answer sheet with a heavy, dark mark so that you cannot see the letter.

1. $\dfrac{3x - 1}{2x + 1}$ is undefined if $x =$

(A) $-\dfrac{1}{2}$

(B) $-\dfrac{1}{3}$

(C) 0

(D) $\dfrac{1}{3}$

(E) $\dfrac{1}{2}$

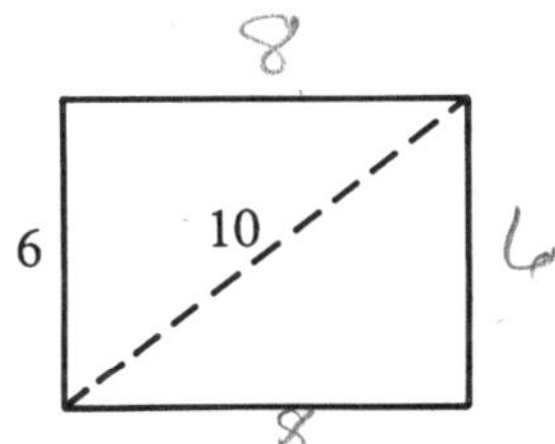

2. In the rectangle above, the perimeter is

(A) 8
(B) 24
(C) 28
(D) 38
(E) 48

3. If 5 percent of x is 2.4, then 500 percent of x is

(A) 12
(B) 24
(C) 120
(D) 240
(E) 2,400

4. If the average distance from the Earth to the Sun is 93,000,000 miles, and the average distance from the Earth to the Moon is 238,000 miles, then the Sun is about how many times as far from the Earth as the Moon is?

(A) 400
(B) 1,800
(C) 4,000
(D) 18,000
(E) 40,000

5. Which of the following fractions is equal to the infinite decimal $0.2\overline{727}$?

(A) $\dfrac{3}{1,100}$

(B) $\dfrac{1}{37}$

(C) $\dfrac{3}{110}$

(D) $\dfrac{27}{100}$

(E) $\dfrac{3}{11}$

6. Which of the following must be equal to zero for all real numbers x ?

I. $-\dfrac{1}{x}$

II. $x + (-x)$

III. x^0

(A) I only
(B) II only
(C) I and III only
(D) II and III only
(E) I, II, and III

7. For $xy \neq 0$, $x^{-1} - y^{-2} =$

(A) $\dfrac{-1}{x} - \dfrac{-2}{y}$

(B) $\dfrac{1}{x - y^2}$

(C) $\dfrac{1}{x} - \dfrac{2}{y}$

(D) $\dfrac{y^2 - x}{xy^2}$

(E) $\dfrac{x - y^2}{xy^2}$

GO ON TO THE NEXT PAGE.

8. A bottling plant has an old machine that fills 6,000 bottles per hour and a new machine that fills 6,000 bottles every 30 minutes. How many minutes will it take both machines working together to fill 6,000 bottles?

(A) 20

(B) 30

(C) 45

(D) $66\frac{2}{3}$

(E) 90

9. If x, $x + 1$, and $x + 2$ are consecutive integers, which of the following must be true?

   I. The average of the three integers is divisible by 2.
   II. The sum of the three integers is divisible by the middle integer.
   III. $(x + 1)(x + 2) - x(x + 1)$ is divisible by 2.

(A) I only
(B) II only
(C) III only
(D) II and III only
(E) I, II, and III

10. For $x \neq 0$, $\dfrac{5x - 3}{3x} + \dfrac{1}{x} =$

(A) $\dfrac{5x - 2}{4x}$

(B) $\dfrac{8x - 3}{3x}$

(C) $\dfrac{5x - 2}{3x}$

(D) $\dfrac{5}{3}$

(E) $-\dfrac{5}{3}$

11. If the area of each face of a cube is 9, what is the volume of the cube?

(A) 18
(B) 27
(C) 54
(D) 81
(E) 729

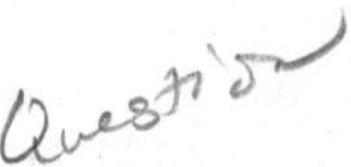

12. If x is a real number and $\sqrt{2x + 7} = x + 2$, then x could be

(A) -3 only
(B) -1 only
(C) 1 only
(D) -1 or -3
(E) -1 or 3

13. Of the following, which is closest to 196 ?

(A) $(8\sqrt{3})^2$
(B) $(6\sqrt{5})^2$
(C) $(5\sqrt{7})^2$
(D) $(9\sqrt{2})^2$
(E) $(5\sqrt{6})^2$

14. A student divides as follows:

$$\frac{15}{32} \div \frac{3}{8} = \frac{15 \div 3}{32 \div 8} = \frac{5}{4}$$

Of the following, which is the best teacher response to the student?

(A) This is an incorrect method for division of fractions; the method works only in this specific problem.
(B) This is an incorrect method for division of fractions; neither of the fractions was inverted.
(C) This method will yield a correct answer only in special cases; the only method that will always yield a correct answer is to invert the second fraction and multiply.
(D) This method always yields an equivalent expression, but not always in its simplest form.
(E) This method will yield a correct answer only if the quotient is greater than 1.

GO ON TO THE NEXT PAGE.

FREQUENCY DISTRIBUTION OF SCORES
ON AN ALGEBRA TEST

Test Score	Frequency
100	1
95	3
90	4
85	4
80	4
75	6
70	4
65	3
60	1
Total Number of Students	30

15. What is the mode of the distribution above?

(A) 70
(B) 75
(C) 79
(D) 80
(E) 82

16. Which of the following statements about measurement is (are) true?

   I. Both metric units and English units are standard units.
  II. Physical measurements are, at best, approximations.
 III. Physical measurements made with nonstandard units can be useful.

(A) II only
(B) I and II only
(C) I and III only
(D) II and III only
(E) I, II, and III

17. Which of the following sources would be most likely to have some suggestions on classroom-tested techniques for motivating students in a high school geometry class?

(A) *Mathematical Reviews*
(B) *Mathematics Teacher*
(C) *The American Mathematical Monthly*
(D) *Arithmetic Teacher*
(E) *Transactions of the American Mathematical Society*

18. If $\log_x 25 = \frac{2}{3}$, then $x =$

(A) -5
(B) 5
(C) 25
(D) 125
(E) 625

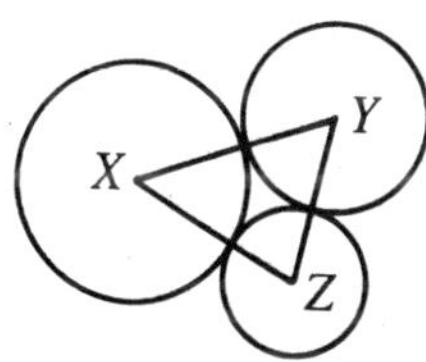

19. In the figure above, X, Y, and Z are the centers of circles that have diameters of 24, 14, and 10, respectively. What are the lengths of the sides of $\triangle XYZ$?

(A) 12, 7, and 5
(B) 19, 17, and 12
(C) 24, 14, and 10
(D) 38, 34, and 24
(E) 48, 28, and 20

20. If $2xy = 2x$, which of the following must be true?

(A) $y = 0$
(B) $x = 0$
(C) $x = y = 0$
(D) $x = y = 1$
(E) $x = 0$ or $y = 1$

21. $2^9 \cdot 5^7$ is an integral multiple of all of the following EXCEPT

(A) 1
(B) 2
(C) 8
(D) 75
(E) 100

GO ON TO THE NEXT PAGE.

22. Which of the following is FALSE for all real numbers?

(A) If $x + y > z$, then $x > z - y$.

(B) If $\frac{x}{y} > z$ and $y > 0$, then $x > yz$.

(C) If $x - y > 0$, then $x > y$.

(D) If $x < y$ and $x > 0$, then $x^2 < y^2$.

(E) If $-3x < 6$, then $x < -2$.

23. Of the following numbers, which is the arithmetic mean of 4 consecutive multiples of 3 ?

(A) 5

(B) $7\frac{1}{2}$

(C) 9

(D) $10\frac{1}{3}$

(E) 12

24. To determine whether a positive integer N is a prime number, it is necessary and sufficient to check whether N is divisible by any prime numbers less than or equal to

(A) N

(B) N^2

(C) $\sqrt{N}$

(D) $\frac{1}{2}N$

(E) $N - 1$

25. If a student states that

$$-\frac{1}{2}\left(-\frac{2}{3}x + \frac{1}{2}\right) = \frac{1}{3}x + \frac{1}{2},$$

it is most likely that the mistake results from a misunderstanding of which of the following?

(A) Multiplication of fractions
(B) Arithmetic of negative numbers
(C) Commutative property of multiplication
(D) Distributive property
(E) Equivalence relation

26. If $A = \begin{pmatrix} 1 & 0 \\ 1 & 1 \end{pmatrix}$ and $B = \begin{pmatrix} 1 & 1 \\ 0 & 1 \end{pmatrix}$, then $AB =$

(A) $\begin{pmatrix} 1 & 1 \\ 1 & 2 \end{pmatrix}$

(B) $\begin{pmatrix} 2 & 1 \\ 1 & 1 \end{pmatrix}$

(C) $\begin{pmatrix} 2 & 1 \\ 1 & 2 \end{pmatrix}$

(D) $\begin{pmatrix} 1 & 0 \\ 0 & 1 \end{pmatrix}$

(E) $\begin{pmatrix} 1 & 1 \\ 1 & 1 \end{pmatrix}$

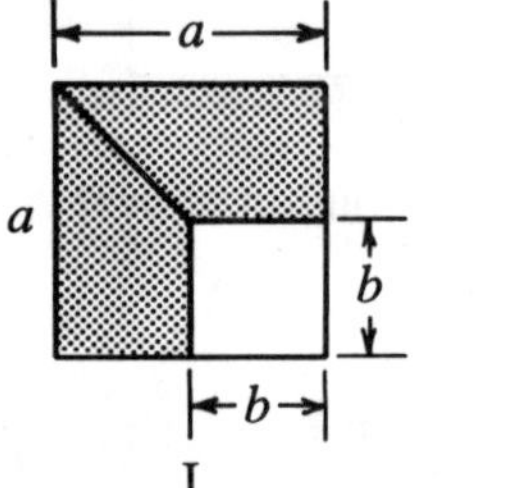
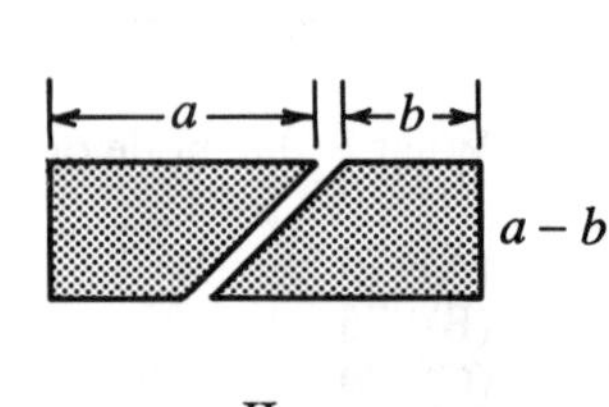

27. In the figures above, the two shaded trapezoidal regions of the square in Figure I can be rearranged as shown in Figure II. This illustration can be used to show students a geometric verification of which of the following identities?

(A) $a^2 + ab = a(a + b)$
(B) $a^2 - ab = a(a - b)$
(C) $a^2 - b^2 = (a + b)(a - b)$
(D) $a^2 - 2ab + b^2 = (a - b)^2$
(E) $a^2 + 2ab + b^2 = (a + b)^2$

GO ON TO THE NEXT PAGE.

$$\begin{cases} x^2 + y^2 = 25 \\ x + y = 0 \end{cases}$$

28. The pair of equations above has how many common solutions?

    (A) None
    (B) Exactly 1
    (C) Exactly 2
    (D) Exactly 4
    (E) Infinitely many

29. When the base of a triangle is decreased by 20 percent and the altitude to this base is increased by 20 percent, the area is

    (A) increased by 4%
    (B) increased by 2%
    (C) unchanged
    (D) decreased by 2%
    (E) decreased by 4%

30. If $\tan(t + k) = \tan t$ for all real numbers t for which $\tan(t + k)$ and $\tan t$ exist, then k could be

    (A) $\dfrac{3\pi}{2}$

    (B) π

    (C) $\dfrac{2\pi}{3}$

    (D) $\dfrac{\pi}{2}$

    (E) $\dfrac{\pi}{4}$

31. If the area of an isosceles right triangle is 32, what is the perimeter of the triangle?

    (A) $12\sqrt{2}$
    (B) $8 + 8\sqrt{2}$
    (C) $16 + 8\sqrt{2}$
    (D) 32
    (E) 64

32. If $r > s > 1$, then $\dfrac{\log r - \log s}{\log\left(\dfrac{r}{s}\right)} - 1$ equals

    (A) 0

    (B) $\dfrac{1}{2}$

    (C) 1

    (D) $\dfrac{r}{s(r - s)}$

    (E) $\dfrac{s(r - s)}{r}$

33. Of the following mathematicians, which one is credited with the creation of coordinate geometry?

    (A) Leonhard Euler
    (B) Leonardo Fibonacci
    (C) Evariste Galois
    (D) René Descartes
    (E) Karl Gauss

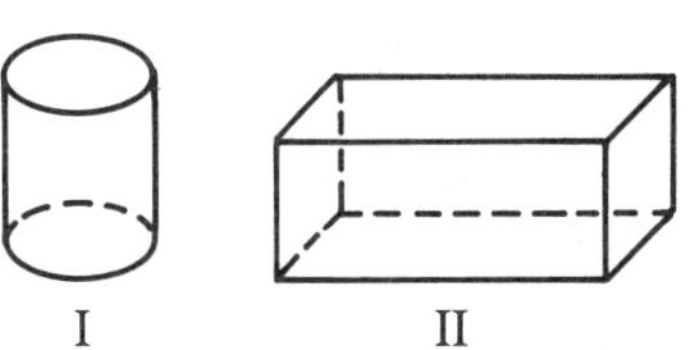

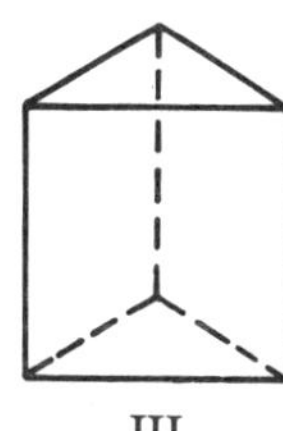
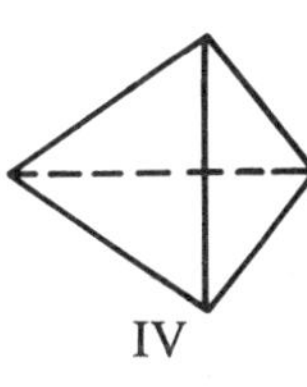

34. The volume(s) of which of the solids shown above can be found by applying the formula $V = Bh$, where V represents the volume, B represents the area of a base, and h represents the altitude to the base that has area B?

    (A) II only
    (B) I and II only
    (C) I and III only
    (D) I, II, and III only
    (E) I, II, III, and IV

GO ON TO THE NEXT PAGE.

35. Two answer sheets are different if at least one
answer on the two sheets is different. How many
different answer sheets are possible for a test
consisting of 10 questions, all of which are
answered true or false?

(A)   20
(B)   64
(C)   100
(D) 1,024
(E) More than 10,000

36. If $40,404 - (4,040,400 \times 10^k) = 0$, then $k =$

(A) -3
(B) -2
(C) -1
(D)   2
(E)   4

37. Which of the following represents the graph of
$$y = \frac{|x + 1|}{x + 1}?$$

(A) 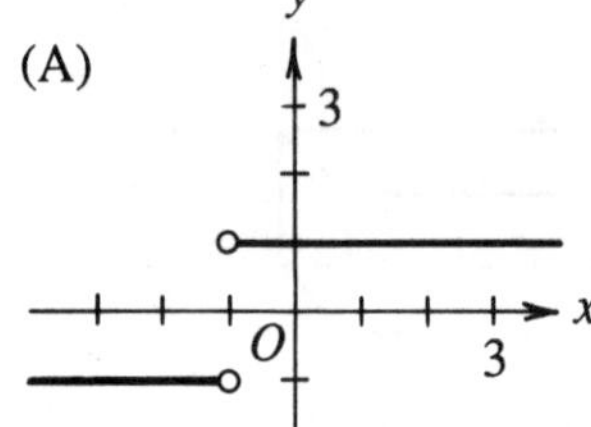

(B) 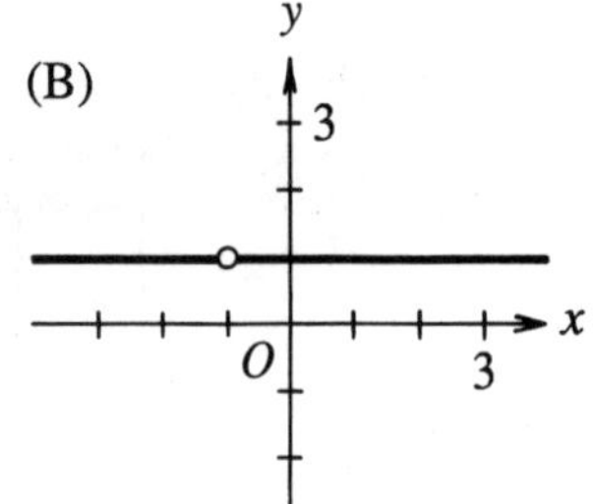

(C) 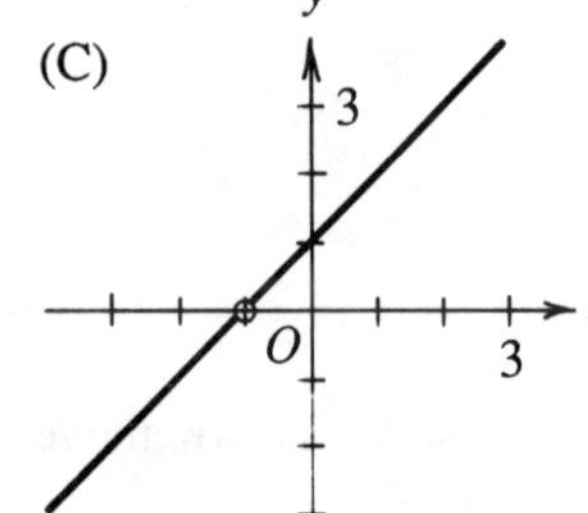

(D) 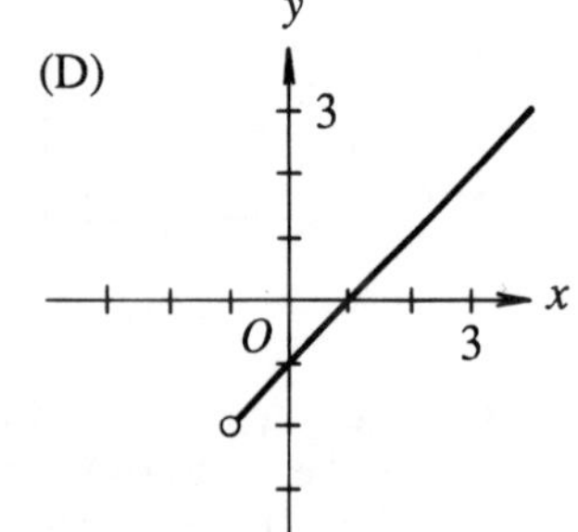

(E) 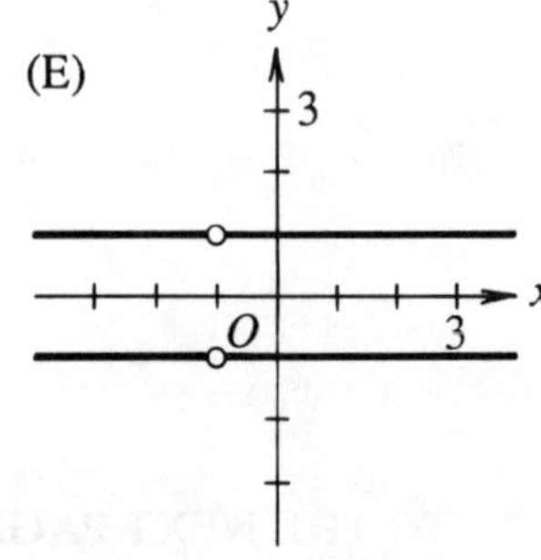

38. Which of the following methods would be appropriate in proving the conjecture, "All positive
integers have property S"?

   I. Finding a nonpositive integer that has
property S
  II. Showing that S is false for any nonpositive
integer
III. Using the principle of mathematical induction

(A) II only
(B) III only
(C) I and II only
(D) I and III only
(E) I, II, and III

39. The equation of the line that is parallel to the line
$x + 3 = y - 2$ and contains the point $(2, -3)$ is

(A) $x + y + 1 = 0$
(B) $x - y - 1 = 0$
(C) $x - y - 5 = 0$
(D) $x + y - 5 = 0$
(E) $x + y + 5 = 0$

40. If x and y are integers and $x + y = 56$, which
of the following must be true of $x - y$?

(A) $x - y$ is odd.
(B) $x - y$ is even.
(C) $x - y < 0$
(D) $x - y > 0$
(E) $x - y < 56$

41. A student concludes that since $\frac{3}{3} = 1$ and $\frac{5}{5} = 1$,
then $\frac{0}{0} = 1$. Which of the following would be the
most appropriate way for the teacher to respond to
this student's conclusion?

(A) To agree and explain that a number divided by
itself is always equal to one
(B) To agree and show that checking the division
by multiplication confirms that $0 \cdot 1 = 0$
(C) To disagree and explain that zero divided by
any number is always equal to zero
(D) To disagree and explain that a unique result
cannot be found since $0 \cdot a = 0$ for any
value of a
(E) To disagree and explain that any number
divided by zero is infinite

GO ON TO THE NEXT PAGE.

42. The initial point of the vector $\overrightarrow{PT}$ is $P(1, 3)$ and the terminal point is $T(4, -2)$. What is the terminal point of a vector that is equivalent to $\overrightarrow{PT}$ and has the origin as its initial point?

(A) $(-3, 5)$
(B) $(1, -3)$
(C) $(3, -5)$
(D) $(5, 1)$
(E) $(5, 5)$

43. Which of the following sets of numbers is closed under the given operation?

(A) The real numbers under division
(B) The irrational numbers under addition
(C) The irrational numbers under multiplication
(D) The odd integers under addition
(E) The odd integers under multiplication

44. If $\displaystyle\sum_{i=1}^{n} i = \frac{n(n+1)}{2}$, then $\displaystyle\sum_{i=1}^{100} 3i =$

(A) $\dfrac{(100)(101)}{2} + 300$

(B) 100^3

(C) $\dfrac{(100)(101)(102)}{6}$

(D) $\dfrac{(300)(301)}{2}$

(E) $3\left[\dfrac{(100)(101)}{2}\right]$

45. The product of complex numbers $(2 + 3i)(3 - 2i)$ equals

(A) 0
(B) 12
(C) $5i$
(D) $6 - 6i$
(E) $12 + 5i$

46. $\dfrac{6!}{3!} =$

(A) $5!$
(B) $4!$
(C) $3(3!)$
(D) 3
(E) 2

47. When a student solves the system of equations $x = 3y$ and $3y = 2x - 1$ by first writing $x = 2x - 1$, the student is using which of the following properties?

(A) Reflexive
(B) Symmetric
(C) Transitive
(D) Commutative
(E) Associative

48. If one assumes that a student A years old requires H hours of sleep per day, and that $H = 14 - \dfrac{A}{3}$, then how old is a student who requires 9 hours of sleep per day?

(A) 18
(B) 15
(C) 11
(D) 9
(E) 8

49. What are all values of x such that $x < \sqrt{x}$?

(A) $0 < x < 1$
(B) $0 < x$
(C) $1 < x$
(D) $x < 0$
(E) $-1 < x < 0$

50. The graph of which of the following equations is a line that is perpendicular to the line represented by the equation $3x + 2y = 5$?

(A) $y = -\dfrac{3}{2}x - 5$

(B) $y = -\dfrac{2}{3}x + 7$

(C) $y = -\dfrac{5}{2}x + \dfrac{3}{2}$

(D) $y = \dfrac{3}{2}x + 7$

(E) $y = \dfrac{2}{3}x + 3$

GO ON TO THE NEXT PAGE.

51. If $f(x) = \sqrt{x^2 + 7}$ for all real numbers x, then $f'(3) =$

(A) $\dfrac{1}{8}$

(B) $\dfrac{1}{4}$

(C) $\dfrac{3}{4}$

(D) 2

(E) 12

52. Circle K and square S each have area π. What is the ratio of the circumference of K to the perimeter of S?

(A) $\dfrac{\sqrt{\pi}}{4}$

(B) $\dfrac{2}{\pi}$

(C) $\dfrac{\pi}{4}$

(D) $\dfrac{\sqrt{\pi}}{2}$

(E) $\dfrac{\pi^2}{16}$

53. Which of the following is NOT a one-to-one function on the set of all real values of x for which the expression is defined?

(A) $f(x) = |x|$
(B) $f(x) = \sqrt{x}$
(C) $f(x) = 2^x$
(D) $f(x) = 2x + 1$
(E) $f(x) = x^3$

54. Which of the following sets of numbers has a greatest element?

(A) The integers
(B) The negative integers
(C) The rational numbers
(D) The negative rational numbers
(E) The negative real numbers

o	a	b	c	d
a	d	c		
b	c			
c			c	
d	b			

55. The table above shows some of the entries in the multiplication table for a group $G = \{a, b, c, d\}$, with multiplication denoted by o. What is the multiplicative inverse of b?

(A) a
(B) b
(C) c
(D) d
(E) It cannot be determined from the information given.

56. At $x = \dfrac{1}{3}$, the graph of $f(x) = x^3 - x^2 - x - 1$ has which of the following?

(A) A relative minimum
(B) A relative maximum
(C) An asymptote
(D) An x-intercept
(E) A point of inflection

57. T is the composite transformation of the coordinate plane that consists of a reflection in the x-axis followed by a reflection in the line through $(0, 0)$ and $(1, 1)$. If (x, y) is an arbitrary point in the plane, what is its image under T?

(A) $(-y, x)$
(B) $(-y, -x)$
(C) $(y, -x)$
(D) $(x, -y)$
(E) (x, y)

GO ON TO THE NEXT PAGE.

58. If $N = \sqrt{(x - 2)^2}$, which of the following must be true?

    I. $N = x - 2$
   II. $N = |x - 2|$
  III. $N \geqq x - 2$

(A) I only
(B) II only
(C) III only
(D) I and III
(E) II and III

59. The point P can be written in polar coordinates as $\left(2, \dfrac{3\pi}{4} \right)$. How is point P written in rectangular coordinates?

(A) $(-2, -2)$

(B) $(-2, 2)$

(C) $(-\sqrt{2}, \sqrt{2})$

(D) $(-\sqrt{2}, -\sqrt{2})$

(E) $(\sqrt{2}, -\sqrt{2})$

60. Of the following, which is most relevant to a classroom discussion of conic sections?

(A) The law of refraction
(B) The orbit of a planet
(C) Ohm's law
(D) Compound interest
(E) Enlargement of a photograph

61. If (m, n) represents the greatest common divisor of the positive integers m and n, which of the following statements is true if $m \neq n$?

(A) $(m, n) > 1$
(B) $(m, n) > m$
(C) $(m, n) < n$
(D) $(m, n) < mn$
(E) None of the above

62. Which of the following can be used to develop the idea of similarity in geometry?

    I. Blueprints of floor plans
   II. Official road maps
  III. Scale models

(A) I only
(B) II only
(C) III only
(D) I and II only
(E) I, II, and III

GO ON TO THE NEXT PAGE.

23

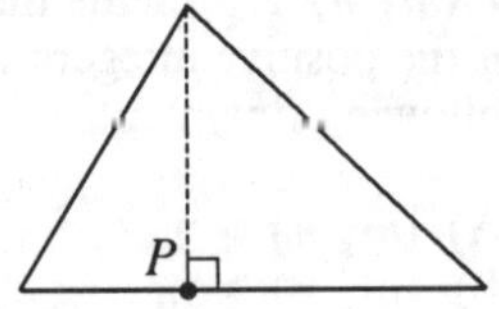
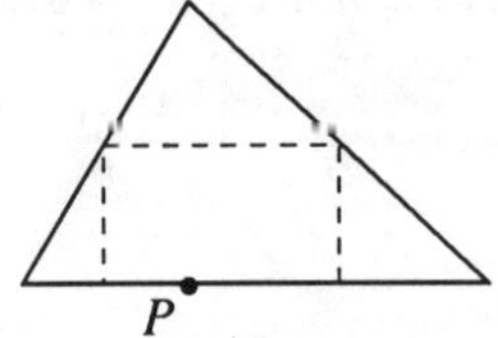
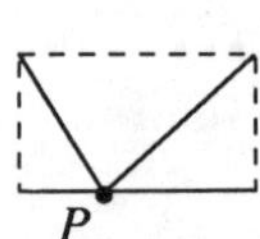

| Figure I | Figure II | Figure III |

63. In a geometry class, the teacher gives each student a triangular piece of paper marked as shown in Figure I. The students are then instructed to fold the triangle so that each vertex coincides with point P and a rectangle is formed by the dotted lines (folds), as shown in Figures II and III. Which of the following could the students discover from this demonstration?

   I. The sum of the measures of the angles of a triangle is equal to the measure of a straight angle.

   II. The length of the segment that joins the mid-points of two sides of a triangle is one-half the length of the third side.

   III. The area of the rectangular region formed as shown in Figure III is half the area of the original triangular region shown in Figure I.

(A) I only
(B) II only
(C) III only
(D) I and II only
(E) I, II, and III

---

64. For what values of x does the determinant

$$\begin{vmatrix} 1 & -x \\ -x & 1 \end{vmatrix} \text{ equal } 0 ?$$

(A) No values of x
(B) $x = -1$ only
(C) $x = 0$
(D) $x = 1$ only
(E) $x = -1$ or $x = 1$

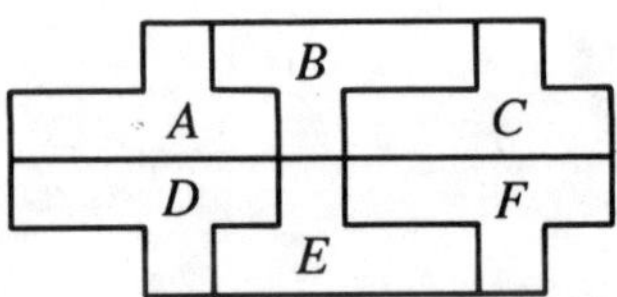

65. The figure above shows six identical tiles, all of which are red on one side and white on the other side. If tile A has the red side turned up, which of the other tiles have the red side up?

(A) B and C
(B) D and E
(C) B and D
(D) C and E
(E) C and F

GO ON TO THE NEXT PAGE.

66. If $\cos x \neq 0$, then $\dfrac{1 - \cos^2 x}{\cos^2 x} =$

(A) $\tan^2 x$

(B) $\sec^2 x$

(C) $\csc^2 x$

(D) $\cot^2 x$

(E) $\cos 2x$

67. For the numbered statements below, the notation $p \leftarrow q$ is defined to mean "replace the value of p by the present value of q."

   1. $x \leftarrow 5,\ y \leftarrow 3$
   2. $z \leftarrow (x + y)$
   3. Record the value of z.
   4. $x \leftarrow (x + y)$
   5. Go to step 2.

The <u>second</u> value of z recorded at step 3 will be

(A) 8
(B) 10
(C) 11
(D) 13
(E) 16

68. If the line $5x - 12y = 60$ intersects the x-axis and the y-axis at points P and Q, what is the length of segment PQ?

(A) 5
(B) 7
(C) 12
(D) 13
(E) 17

69. Which of the following describes the roots of $-\dfrac{2x^2}{3} + \dfrac{5x}{7} + \dfrac{7}{10} = 0$?

(A) There is only one root.
(B) One is positive and one is negative.
(C) Both are positive.
(D) Both are negative.
(E) Both are imaginary.

70. Which of the following graphs represents the solution set of $x^2 - 9 < 0$?

(A) 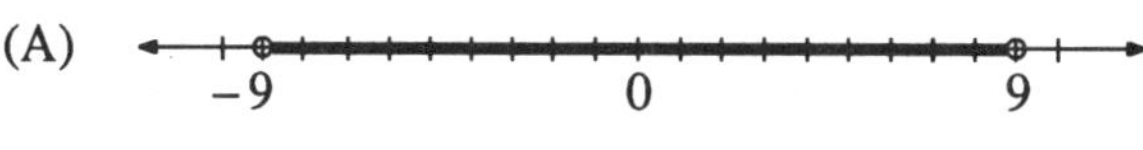

(B) 

(C)

(D) 

(E)

A cyclist traveled from Caldonia to Newtown at an average rate of 18 miles per hour and returned immediately by the same route at an average rate of 12 miles per hour. What was the average rate for the round trip?

71. With respect to the problem above, which of the following is true?

(A) The answer is 14.4 miles per hour.
(B) The answer is 15 miles per hour.
(C) The only additional information needed to determine the answer is the time for the round trip.
(D) The only additional information needed to determine the answer is the distance between the towns.
(E) Both the total time and the distance are needed to determine the answer.

72. A plane intersects a right circular cylinder (with two bases) and is parallel to the axis of the cylinder. If the plane passes through the interior of the cylinder, then the intersection of the plane with the surface of the cylinder is

(A) exactly 4 points
(B) a pair of parallel line segments
(C) a rectangle
(D) a circle
(E) an ellipse

GO ON TO THE NEXT PAGE.

73. If f is a function such that $f(3) = 2$, $f(4) = 2$, and $f(n + 4) = f(n + 3)f(n + 2)$ for all integers $n \geq 0$, what is the value of $f(6)$?

(A) 4
(B) 5
(C) 6
(D) 8
(E) It cannot be determined from the information given.

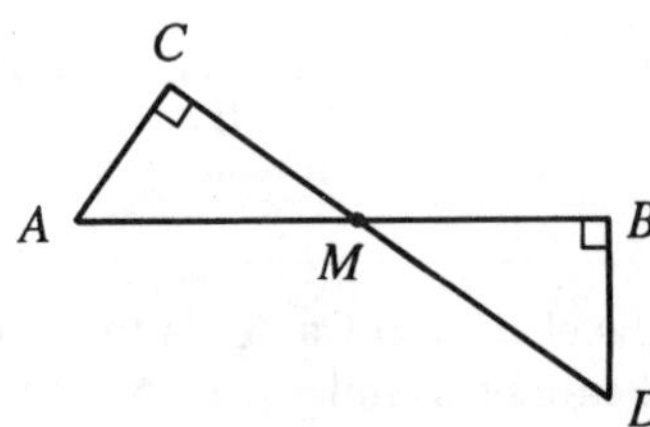

74. In the figure above, if $\overline{AB}$ and $\overline{CD}$ intersect at M, and if $\angle B$ and $\angle C$ are right angles, then $\dfrac{MB}{BD} =$

(A) $\dfrac{MC}{CA}$

(B) $\dfrac{MA}{AC}$

(C) $\dfrac{MC}{MA}$

(D) $\dfrac{MD}{MA}$

(E) $\dfrac{MA}{MC}$

75. If each of the acute angles of a parallelogram has degree measure 60, and the lengths of the sides are 3 and 4, then the area of the parallelogram is

(A) 6
(B) $4\sqrt{3}$
(C) $4\sqrt{6}$
(D) $6\sqrt{3}$
(E) 12

76. If $f(x) = \dfrac{2x - 5}{2}$ and $g(x) = \dfrac{2}{2x - 5}$, then $g(f(10)) =$

(A) $\dfrac{1}{10}$

(B) $\dfrac{1}{5}$

(C) 1

(D) 5

(E) 10

77. S is the set of all numbers which can be written in the form $2^n 3^m$ where n and m are positive integers. If a and b are two numbers in S, which of the following must also be in S?

(A) $a + b$

(B) $a - b$

(C) $\sqrt{ab}$

(D) $\dfrac{a}{b}$

(E) ab

78. Each of the following defines a function such that the interval $[0, 1]$ is mapped into the interval $[0, 1]$ EXCEPT

(A) $f(x) = \sqrt{x}$

(B) $f(x) = |x|$

(C) $f(x) = \dfrac{1}{x}$

(D) $f(x) = x^2$

(E) $f(x) = x^3$

GO ON TO THE NEXT PAGE.

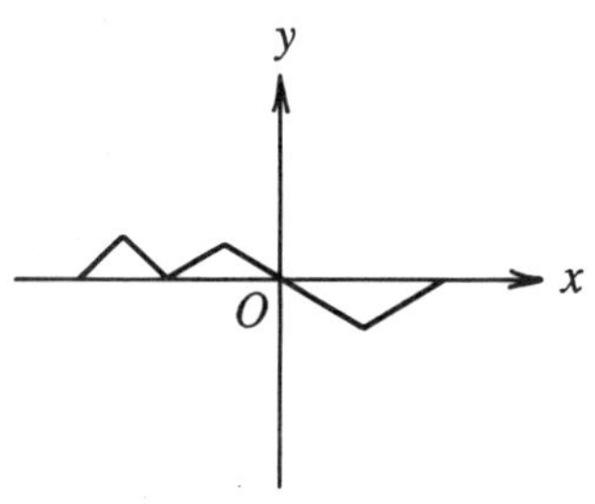

79. The figure above shows the graph of $y = f(x)$.
Which of the following is the graph of $y = -f(x)$?

(A) 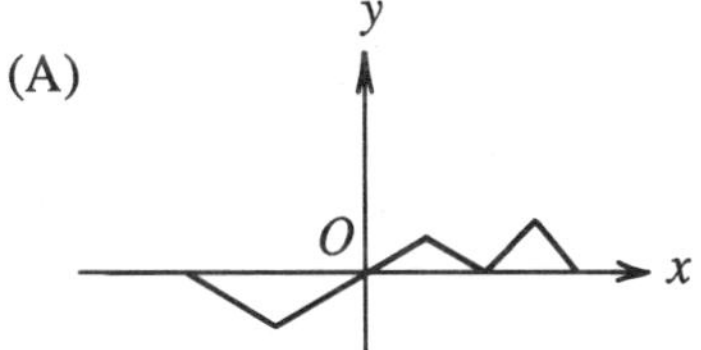

(B) 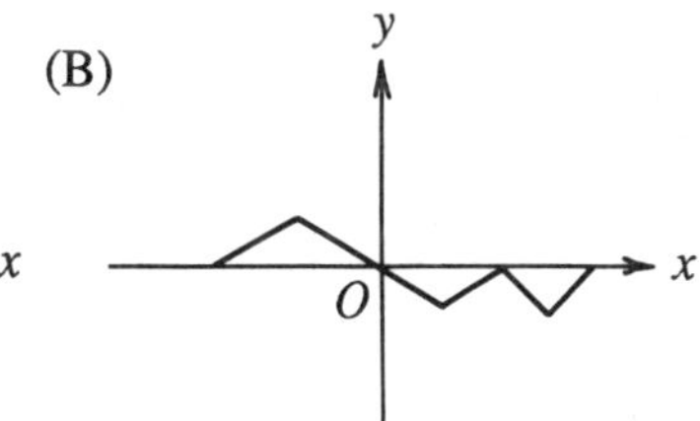

(C) 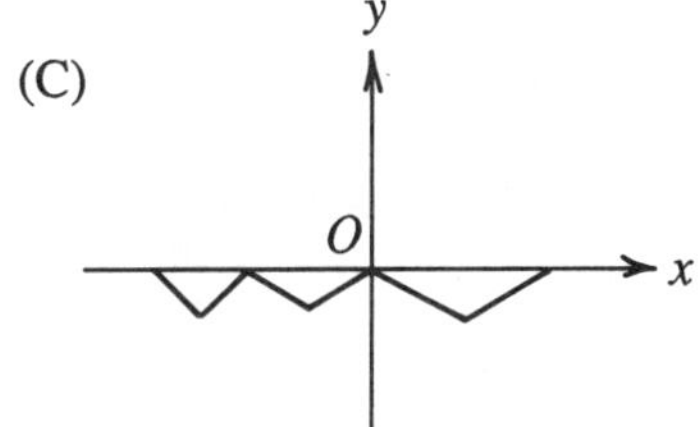

(D) 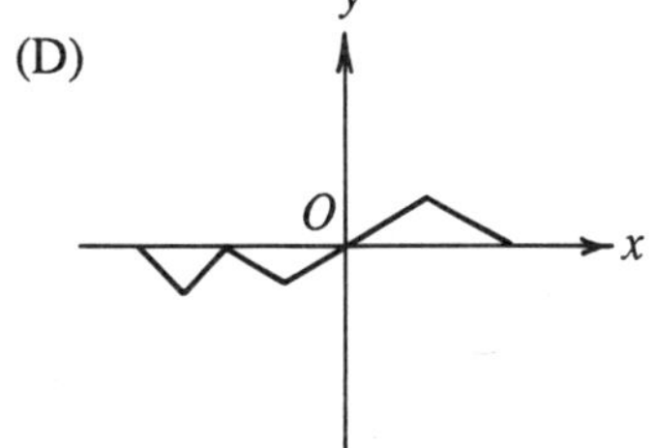

(E) 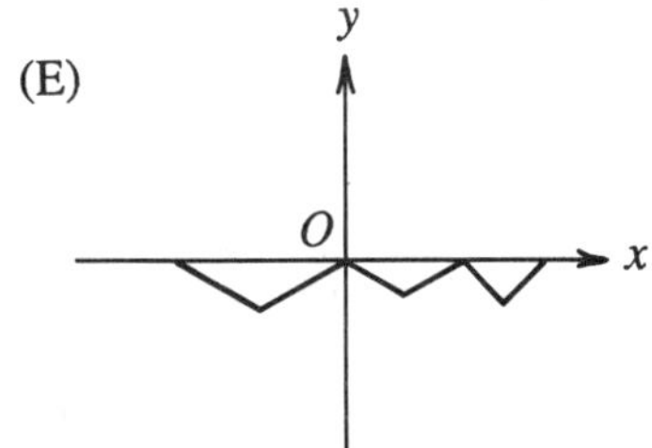

GO ON TO THE NEXT PAGE.

80. If the results of tossing a fair coin 5 times are 5 heads in a row, what is the probability of getting a head on the 6th toss?

(A) 0

(B) $\left(\dfrac{1}{2}\right)^6$

(C) $\dfrac{1}{6}$

(D) $\dfrac{1}{2}$

(E) 1

81. What is the area of the region bounded below by the graph of $y = x^2 - 4$ and bounded above by the graph of $y = 4 - x^2$?

(A) 0

(B) $\dfrac{16}{3}$

(C) $\dfrac{32}{3}$

(D) $\dfrac{64}{3}$

(E) $\dfrac{128}{3}$

82. If length L is measured to be 7.24 meters and if this measurement is accurate to the nearest hundredth of a meter, then the greatest possible error in this measurement is

(A) 0.0005 m
(B) 0.005 m
(C) 0.05 m
(D) 0.5 m
(E) 1.0 m

83. In trying to prove that $\sqrt{3}$ is irrational using the indirect method of proof, one should do which of the following?

(A) Assume that $\sqrt{3}$ is rational and show that this leads to the contradiction of a known fact.

(B) Assume that $\sqrt{3}$ is rational and show that this does not lead to the contradiction of a known fact.

(C) Assume that $\sqrt{3}$ is not rational and show that this leads to the contradiction of a known fact.

(D) Assume that $\sqrt{3}$ is not rational and show that this does not lead to the contradiction of a known fact.

(E) None of the above, since one cannot prove by the indirect method that $\sqrt{3}$ is irrational.

84. The complex numbers satisfy all of the following EXCEPT

(A) the order properties
(B) closure under raising to a power
(C) associativity under multiplication
(D) the group properties for addition
(E) the distributive property

85. Let G be a group with operation $*$. If a, b, and c are elements in G, then which of the following statements is NOT necessarily true?

(A) $a * b$ is in G
(B) $a * (b * c) = (a * b) * c$
(C) $a * b = b * a$
(D) There exists an element e in G such that $a * e = e * a = a$ for each a in G.
(E) For each a in G there exists an element a^{-1} in G such that $a * a^{-1} = e$.

GO ON TO THE NEXT PAGE.

86. If f is a trigonometric function whose graph has positive slope wherever it is defined, then $f(x) =$

(A) sin x
(B) cos x
(C) tan x
(D) sec x
(E) cot x

87. What are all possible values of x for which $\cos^2 x = \cos x$ and $0 \leq x < 2\pi$?

(A) 0 only

(B) $\dfrac{3\pi}{2}$ only

(C) 0 and $\dfrac{\pi}{2}$ only

(D) $\dfrac{\pi}{2}$ and $\dfrac{3\pi}{2}$ only

(E) 0, $\dfrac{\pi}{2}$, and $\dfrac{3\pi}{2}$

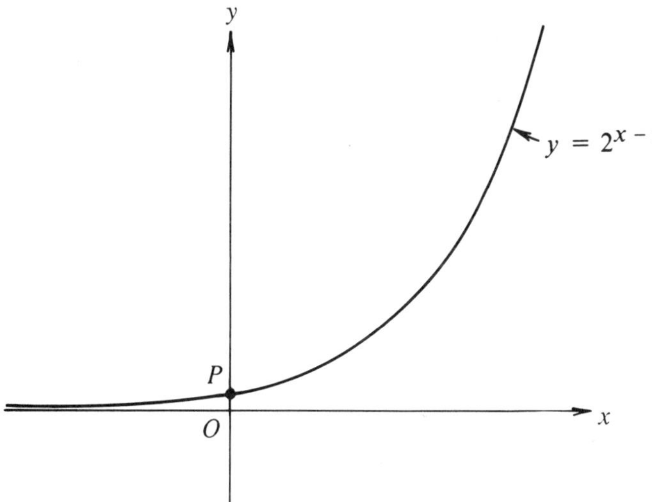

88. The figure above shows the graph of $y = 2^{x-2}$. The point P is

(A) $\left(\dfrac{1}{4},\, 0\right)$

(B) $(1, 0)$

(C) $(0, 1)$

(D) $\left(0,\, \dfrac{1}{2}\right)$

(E) $\left(0,\, \dfrac{1}{4}\right)$

89. Which of the following do NOT necessarily determine a plane?

(A) A line and a point not on the line
(B) Two distinct parallel lines
(C) The vertices of a triangle
(D) Two distinct intersecting lines
(E) Three distinct points

90. If y varies directly as x^2, and if $y = 1$ when $x = 0.1$, for what values of x does $y = \dfrac{1}{2}$?

(A) $\pm\dfrac{1}{200}$

(B) $\pm\dfrac{1}{20}$

(C) $\pm\dfrac{\sqrt{2}}{20}$

(D) $\pm\sqrt{5}$

(E) $\pm 5\sqrt{2}$

91. Let f be the function given by $f(x) = 2x - 1$. The inverse function of f is given by $f^{-1}(x) =$

(A) $1 - 2x$

(B) $\dfrac{1}{2x - 1}$

(C) $\dfrac{x + 1}{2}$

(D) $2x + 1$

(E) $\dfrac{x - 1}{2}$

GO ON TO THE NEXT PAGE.

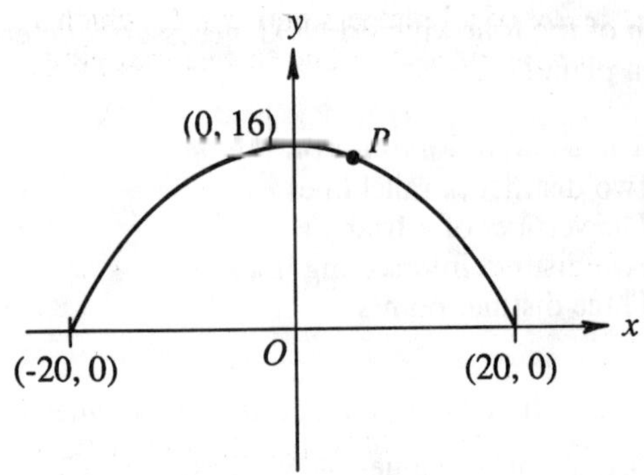

92. The figure above indicates the coordinates of three points on the parabolic arc. If P is the point on the arc that has coordinates $(5, y)$, then $y =$

(A) 12

(B) 15

(C) $15\frac{1}{3}$

(D) $15\frac{1}{2}$

(E) $15\frac{3}{4}$

$$\begin{cases} 3^{x+y} = 81 \\ 36^{\frac{y}{2}} = 216 \end{cases}$$

93. In the system of equations above, $x =$

(A) 0
(B) 1
(C) 2
(D) 3
(E) 4

94. The slope of a curve at every point (x, y) on the curve is given by $10x - 1$. If the curve contains the point $(1, 5)$, then an equation of the curve is

(A) $y = 10x^2 + x - 1$
(B) $y = 5x^2 + x + 1$
(C) $y = 5x^2 - x$
(D) $y = 5x^2 - x + 1$
(E) $y = 5x^2 + x$

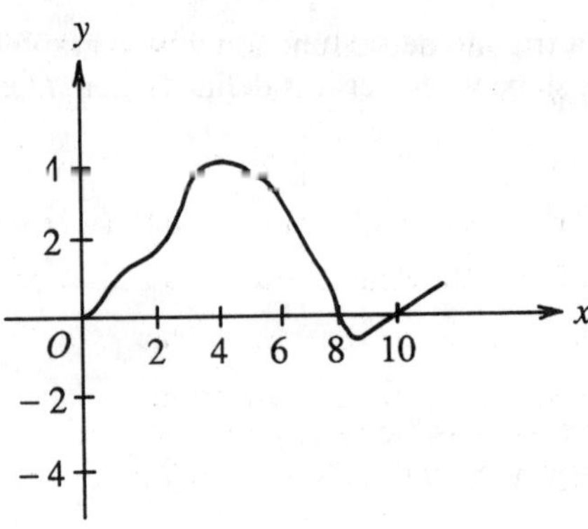

95. For all x in the interval $[0, 10]$, F is the function defined by $F(x) = \int_0^x f$, where f is the function that has the graph shown above. At what value of x does F attain its maximum value?

(A) 0
(B) 4
(C) 6
(D) 8
(E) 10

96. Which of the following could NOT be used as a definition of the Euclidean distance between any two distinct points x_1 and x_2 on the number line?

(A) $\begin{cases} x_2 - x_1, & \text{if } x_2 > x_1 \\ x_1 - x_2, & \text{if } x_1 > x_2 \end{cases}$

(B) $\left| \int_{x_1}^{x_2} dx \right|$

(C) $\sqrt{(x_1 - x_2)^2}$

(D) $x_1 - x_2$

(E) $\max \{(x_1 - x_2), (x_2 - x_1)\}$

GO ON TO THE NEXT PAGE.

97. A teacher instructed a student to factor completely the polynomial $x^3 + 2x^2 - 5x - 10$. The teacher expected the answer $(x^2 - 5)(x + 2)$, but the student wrote $(x + \sqrt{5})(x - \sqrt{5})(x + 2)$. Which of the following instructions would have indicated more precisely what the teacher expected?

   I. Factor into a product of primes.
   II. Factor completely over the integers.
   III. Factor into polynomials that have integer coefficients and are of the lowest possible degree.

(A) I only
(B) II only
(C) III only
(D) I or III
(E) II or III

98. An algorithm for multiplying decimals is as follows:

   1. Treat the numbers as whole numbers and multiply.
   2. In the result, point off from right to left the total number of decimal places in the two factors.

This algorithm is based on which of the following?

(A) $\dfrac{a^n}{10^n} \times \dfrac{b^n}{10^n} = \dfrac{(ab)^n}{10^{2n}}$

(B) $\dfrac{a}{10^m} \times \dfrac{b}{10^n} = \dfrac{ab}{10^{m+n}}$

(C) $a^n(b^n + c^n) = (ab)^n + (ac)^n$

(D) If $ab = c$, then $\dfrac{c}{a} = b$.

(E) If $\dfrac{c}{a} = b$, then $ab = c$.

99. If a box of 2 dozen ballpoint pens contains 5 that are defective, what is the probability that 2 pens both selected at random and without replacement will be defective?

(A) $\left(\dfrac{5}{24}\right)\left(\dfrac{4}{23}\right)$

(B) $\left(\dfrac{5}{24}\right)\left(\dfrac{4}{24}\right)$

(C) $\left(\dfrac{19}{24}\right)\left(\dfrac{19}{23}\right)$

(D) $\left(\dfrac{5}{24}\right)^2$

(E) $\left(\dfrac{19}{24}\right)^2$

100. If I is the set of all integers and $x \in I$, which of the following denotes a one-to-one mapping of I <u>onto</u> I?

(A) $x \to x - 4$
(B) $x \to 2x + 1$
(C) $x \to 2x$
(D) $x \to x^2$
(E) $x \to x^3$

101. The statement, "You will be healthy only if you exercise regularly," is equivalent to which of the following?

   I. If you exercise regularly, then you will be healthy.
   II. If you are healthy, then you exercise regularly.
   III. A sufficient condition for being healthy is that you exercise regularly.

(A) I only
(B) II only
(C) I and II only
(D) I and III only
(E) I, II, and III

102. What is $\lim\limits_{x \to 4} \dfrac{x^3 - 64}{x^2 - 16}$?

(A) 0
(B) 1
(C) 6
(D) 12
(E) The limit does not exist.

103. For which of the following topics would a hand calculator be LEAST useful as a teaching device?

(A) Finding the limit of a function
(B) Finding the length of an unknown side of a triangle by the use of trigonometry
(C) Finding the coordinates of points that satisfy an equation
(D) Finding the mean of a frequency distribution
(E) Proving a theorem by mathematical induction

GO ON TO THE NEXT PAGE.

104. If right circular cylinder A has a radius twice that of right circular cylinder B and a height 3 times that of B, then the volume of A is how many times the volume of B?

(A) 2
(B) 3
(C) 5
(D) 6
(E) 12

105. Which of the following series converges?

(A) $\frac{1}{2} + \frac{2}{3} + \frac{3}{4} + \frac{4}{5} + \frac{5}{6} + \ldots + \frac{n}{n+1} + \ldots$

(B) $1 + \frac{2}{\sqrt{2}} + \frac{3}{\sqrt{3}} + \frac{4}{\sqrt{4}} + \frac{5}{\sqrt{5}} + \ldots + \frac{n}{\sqrt{n}} + \ldots$

(C) $\frac{2}{2} + \frac{2}{3} + \frac{2}{4} + \frac{2}{5} + \ldots + \frac{2}{n+1} + \ldots$

(D) $1 + \frac{1}{2} + \frac{1}{3} + \frac{1}{4} + \ldots + \frac{1}{n} + \ldots$

(E) $1 + \frac{1}{3} + \frac{1}{9} + \frac{1}{27} + \frac{1}{81} + \ldots + \frac{1}{3^{n-1}} + \ldots$

106. If the sequence a_n is defined by

$$a_n = \left[2 + \frac{(-1)^n}{n} \right], \text{ then } \lim_{n \to \infty} a_n \text{ is}$$

(A) $-\infty$
(B) -2
(C) 0
(D) 2
(E) ∞

107. If f is a continuous function on $[a, b]$ and

$f(x) > 0$ for all x in $[a, b]$, then $\int_a^b f(x)dx$

represents the

(A) length of the curve $y = f(x)$ between $x = a$ and $x = b$
(B) area of the region bounded by the graphs of $y = f(x)$, $x = a$, $x = b$, and the x-axis
(C) area of the region bounded by the graphs of $y = f(x)$, $y = a$, $y = b$, and the y-axis
(D) volume of the solid generated by revolving about the x-axis the region bounded by the graphs of $y = f(x)$, $x = a$, $x = b$, and the x-axis
(E) volume of the solid generated by revolving about the y-axis the region bounded by the graphs of $y = f(x)$, $x = a$, $x = b$, and the x-axis

108. The graph of the equation
$x^2 + 4y^2 + 4x - 24y + 36 = 0$ is

(A) a circle
(B) a parabola
(C) an ellipse
(D) a hyperbola
(E) a point

109. The automobile license numbers assigned in a certain county consist of 2 letters followed by 3 digits, and any letter or digit may be repeated. If the letters I and O are excluded, and the first digit cannot be zero, how many different license numbers can be assigned?

(A) $24 \cdot 24 \cdot 9 \cdot 10 \cdot 10$
(B) $26 \cdot 25 \cdot 9 \cdot 9 \cdot 9$
(C) $26 \cdot 25 \cdot 10 \cdot 9 \cdot 8$
(D) $24 \cdot 24 \cdot 9 \cdot 9 \cdot 9$
(E) $24 \cdot 23 \cdot 9 \cdot 8 \cdot 7$

110. If f and g are inverse functions, which of the following is NOT necessarily true?

(A) $f(x) \cdot g(x) = 1$
(B) $f(g(x)) = x$
(C) $f(g(x)) = g(f(x))$
(D) Both f and g are one-to-one functions.
(E) The domain of g is the range of f.

GO ON TO THE NEXT PAGE.

111. If $m < 0$ and $n > 0$, which of the following could be the graph of $y = mx^3 + n$?

(A)
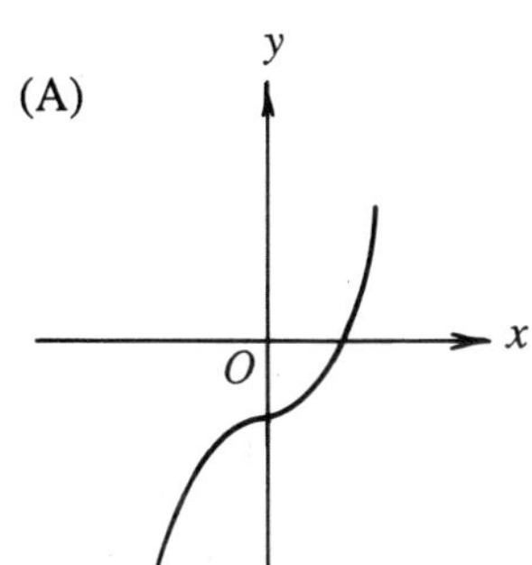

(B)
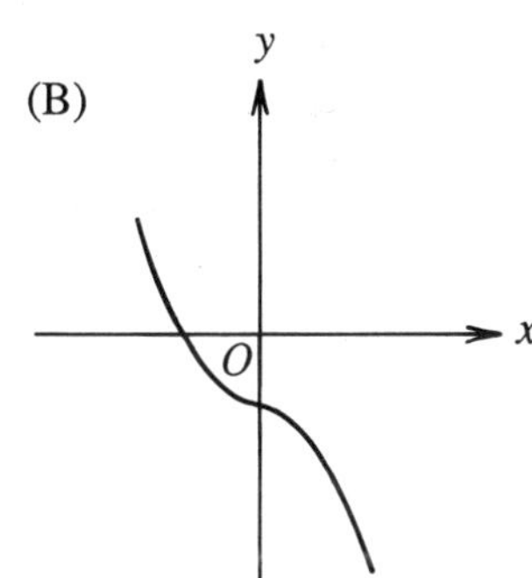

(C)
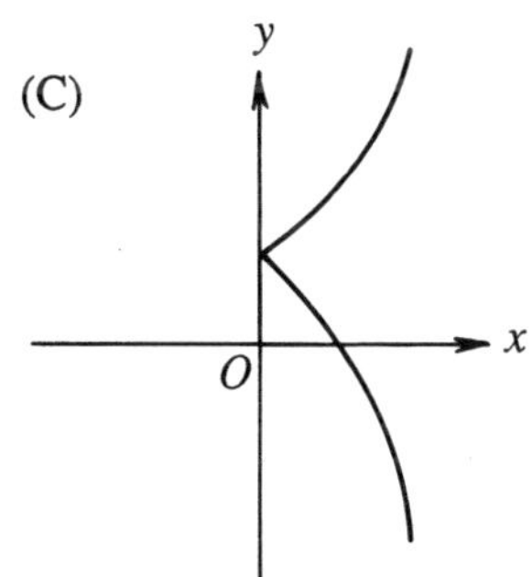

(D)
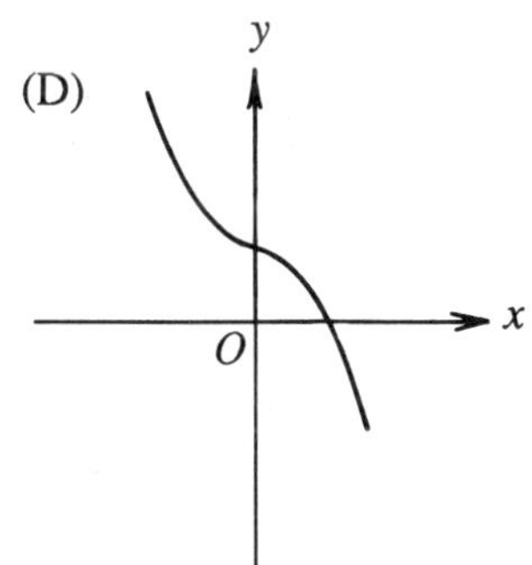

(E)
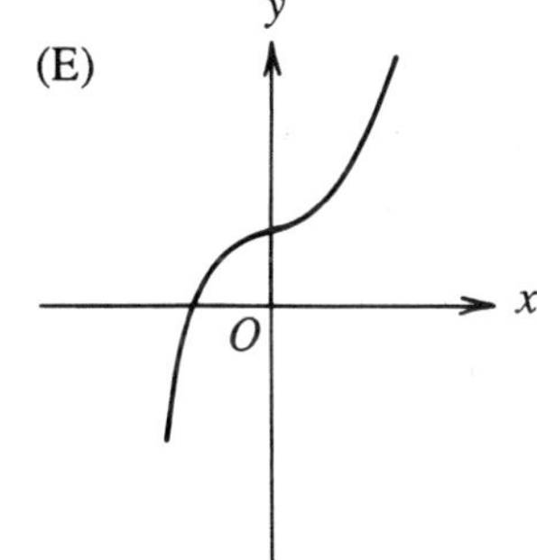

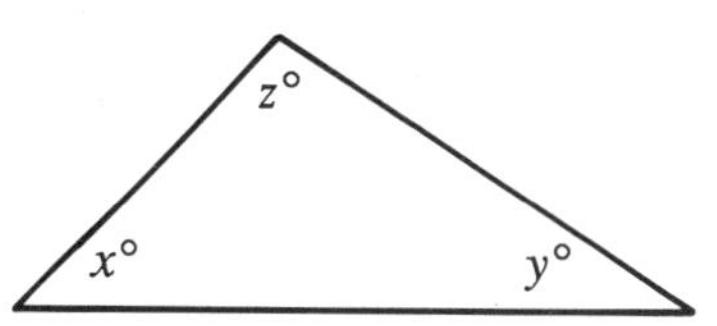

Note: Figure not drawn to scale.

112. In the triangle above, if $60 < y < 110$, then

(A) $60 < x < 110$
(B) $70 < x < 110$
(C) $60 < x - z < 110$
(D) $70 < x + z < 120$
(E) $70 < x - z < 120$

113. The amount to which \$1,000 will grow in 5 years at a 6 percent annual interest rate compounded semi-annually is given by

(A) $1,000(1 + 0.03)^{10}$
(B) $1,000(1 + 0.03)^{5}$
(C) $1,000(1 + 0.06)^{10}$
(D) $1,000(1 + 0.06)^{5}$
(E) $1,000(1 + 0.12)^{5}$

	Set	Relation
A	Integers	is a multiple of
B	Real numbers	is greater than or equal to
C	All sets of integers	is a subset of
D	Triangles	is similar to
E	Lines	is perpendicular to

114. Which of the five relations listed above represents an equivalence relation?

(A) A
(B) B
(C) C
(D) D
(E) E

GO ON TO THE NEXT PAGE.

115. From a point 1,000 meters from the base of an airport tower, the angle of elevation to the top of the tower is 15 degrees. If h is the height, in meters, of the tower, then $h =$

(A) $1{,}000 \tan 15°$

(B) $\dfrac{1{,}000}{\tan 15°}$

(C) $1{,}000 \sin 15°$

(D) $\dfrac{1{,}000}{\sin 15°}$

(E) $1{,}000 \cos 15°$

116. Recommendations made by the National Council of Teachers of Mathematics in their publication *Curriculum and Evaluation Standards for School Mathematics* include which of the following?

   I. The mathematics curriculum should emphasize problem-solving.
   II. All students should have access to calculators throughout their school mathematics program.
   III. There should be increased emphasis on drill with numbers apart from problem contexts.

(A) I only
(B) II only
(C) III only
(D) I and II
(E) II and III

117. What is the fifth term in the binomial expansion of $(x + 1)^8$ when the powers of x are written in descending order?

(A) $3{,}024x^4$
(B) $70x^4$
(C) $56x^5$
(D) $35x^4$
(E) $21x^5$

118. A pack of cards is numbered from 1 to 36 inclusive. If one card is selected at random, what is the probability that the number of the card is a multiple of 4 or 6 but not a multiple of 12?

(A) $\dfrac{7}{12}$

(B) $\dfrac{1}{2}$

(C) $\dfrac{1}{3}$

(D) $\dfrac{1}{4}$

(E) $\dfrac{1}{6}$

119. The scores of 500 students on an examination were normally distributed with a mean of 70 and a standard deviation of 10. Letter grades were assigned as follows:

    Above 86 - A
    81-86 - B
    60-80 - C
    53-59 - D
   Below 53 - F

Approximately how many students received a grade higher than C?

(A)   50
(B)   80
(C) 100
(D) 120
(E) 160

120. If $y = x^2$ and $x^y \cdot y^x = x^w$ for $x > 0$, then, in terms of x, $w =$

(A) $x^2 + 2x$
(B) $x^2 + x + 2$
(C) $x^2 + 2^x$
(D) $2x^2$
(E) $2x^3$

S T O P

IF YOU FINISH BEFORE TIME IS CALLED, YOU MAY CHECK YOUR WORK ON THIS TEST.

MATHEMATICS

The supervisor will tell you when to begin work on the test and when to stop. If you finish the test before time is called, go back and check your work on it.

SHOULD YOU GUESS? In this test your score is based on the number of questions you answer correctly; therefore skipped and wrong answers will not count against you. Work as rapidly as you can without sacrificing accuracy. Do not spend too much time puzzling over a question that seems too difficult for you. Answer the easier questions first, then return to the harder ones. Try to answer every question even if you have to guess.

Where necessary, you may use blank spaces in the test book for scratch paper. Do not use any other paper or the margins or back of the answer sheet to do scratchwork.

YOU ARE TO INDICATE ALL OF YOUR ANSWERS ON THE SEPARATE ANSWER SHEET. No credit will be given for anything written in this examination book. After you have decided which of the suggested answers is best, fill in the corresponding space on the answer sheet. BE SURE THAT EACH MARK IS HEAVY AND DARK AND COMPLETELY FILLS THE ANSWER SPACE. Light or partial marks may not be read by the scoring machine. Give only one answer to each question. If you change an answer, be sure that the previous mark is erased completely. Incomplete erasures may be read as intended answers.

This test may include one or more questions that do not count toward your score.

Time—120 minutes

120 Questions

DO NOT BREAK THE SEAL UNTIL YOU ARE TOLD TO DO SO.

NTE® PROGRAMS
SPECIALTY AREA TEST

USE ONLY A SOFT LEAD PENCIL FOR COMPLETING THIS ANSWER SHEET. ERASE NEATLY AND COMPLETELY.

SIDE 1

1. NAME Omit spaces, hyphens, apostrophes, and Jr., III, etc.

Last (family) Name-first 12 letters

First (given) Name-first 8 letters

M.I.

2. SEX

Male ○

Female ○

3. YOUR MAILING ADDRESS

- Fit address into boxes provided.
- Indicate a space in address by leaving a blank box and filling in the corresponding diamond.

P.O. Box or Street Address

City

State or Province

3a. Zip/Postal Code

Please enter code and fill in the corresponding ovals.

BACKGROUND INFORMATION—Answers to questions 4a and 4b will be used for research purposes only. In no case will an individual's responses be singled out.

4a. Is English the dominant language in your household?

○ Yes   ○ No

4b. Do you understand English as well as or better than any other language?

○ Yes   ○ No

5. How do you describe yourself?

Fill in only one oval.

○ Black, Afro-American, or Negro

○ Mexican American or Chicano

○ Native American, Alaska Native, or Aleut

○ Pacific/Asian American

○ Puerto Rican

○ Other Hispanic or Latin American

○ White

○ Other

6. What is your cumulative undergraduate average to date (based on a system where 4.0 = A)?

○ 3.5 – 4.0

○ 3.0 – 3.49

○ 2.5 – 2.99

○ 2.0 – 2.49

○ 1.5 – 1.99

○ below 1.5

7. Major Field Codes

Use your Critical Information Form to complete this item. (If you do not have a Critical Information Form, leave this item blank.)

MAJOR FIELD CODES

UNDERGRADUATE

GRADUATE

8. DATE OF BIRTH

MONTH | DAY | YEAR

9. SOCIAL SECURITY NUMBER (optional)

10. EDUCATIONAL LEVEL

What is the highest educational level you have reached?

Fill in only one oval.

○ Freshman

○ Sophomore

○ Junior

○ Senior

○ Hold bachelor's degree

○ Enrolled in graduate school

○ Hold master's degree

○ Hold doctoral degree

Q1888-10   I.N. 274300   510VV60P200

NTE® PROGRAMS
SPECIALTY AREA TEST

SIDE 2

11. REGISTRATION NUMBER

12. TEST CENTER NUMBER AND LOCATION

PLEASE PRINT:

Center Name

City

State

13. TODAY'S DATE

MO / DAY / YR

14. COLLEGE WHERE YOU RECEIVED TRAINING RELEVANT TO THE TEST

CODE

ETS USE ONLY

Use your completed Critical Information Form for Items 14, 15, and 16. If you do not have a code number for one or more of these items, leave the item(s) blank. Sealed score reports (code 1000) are no longer available. Scores will be sent only to agencies that are authorized score recipients.

15. AGENCIES TO RECEIVE SCORE REPORTS

CODE — NUMBER 1

CODE — NUMBER 2

CODE — NUMBER 3

ETS USE ONLY

16. STATE/AGENCY CODE

CODE

17. YOUR MAILING ADDRESS PRINT COMPLETE ADDRESS BELOW.

Number and Street

Apt. (if any)

City

State/Province

Zip/Postal Code

Telephone Number: (   )

18. SIGNATURE AND STATEMENT

Please write the following statement in longhand on the lines below.

"I am the person whose name and address appear on this answer sheet."

All NTE tests are confidential. I agree not to disclose to anyone the contents of any NTE test I take, and I accept the conditions set forth in the *Bulletin of Information* concerning the administration of the tests and the reporting of scores.

Sign your full first name, middle name or initial, and last name.

19. TEST BOOK INFORMATION

SERIAL NUMBER

TEST FORM

TEST NAME

20. TEST CODE

BE SURE EACH MARK COMPLETELY FILLS THE INTENDED SPACE AND DOES NOT EXTEND INTO THE ADJACENT OVALS. DO NOT MAKE ANY STRAY MARKS. MAKE ALL ERASURES COMPLETE.

(Answer grid: items 1–200, each with ovals A B C D E)

PART FOUR: SCORING THE PRACTICE TEST

How to Score the Mathematics Practice Test

Compare your answers with the correct answers in Table 1 below. Count the number of questions you answered correctly. Use Table 2 on page 40 to find the scaled score corresponding to the number answered correctly.

Table 1: Mathematics Test

Answers to Practice Test Questions and Percentages of Examinees Choosing Correct Answers

Question Number	Correct Answer	Content* Category	Percent of Examinees Choosing Correct Answer	Question Number	Correct Answer	Content* Category	Percent of Examinees Choosing Correct Answer	Question Number	Correct Answer	Content* Category	Percent of Examinees Choosing Correct Answer
1	A	I	91.6%	41	D	IV	44.6%	81	D	III	20.8%
2	C	I	75.4	42	C	II	64.0	82	B	I	61.0
3	D	I	85.2	43	E	III	43.6	83	A	III	68.1
4	A	I	76.0	44	E	II	74.0	84	A	III	22.1
5	E	I	75.2	45	E	II	65.3	85	C	III	29.2
6	B	I	88.7	46	A	III	81.7	86	C	II	54.9
7	D	I	73.8	47	C	I	72.6	87	E	II	35.0
8	A	I	77.6	48	B	I	89.3	88	E	II	70.1
9	D	I	54.5	49	A	I	65.5	89	E	II	29.0
10	D	I	89.7	50	E	II	55.6	90	C	I	28.2
11	B	I	72.6	51	C	III	57.0	91	C	II	33.4
12	C	I	77.6	52	D	I	45.1	92	B	II	30.2
13	A	I	75.5	53	A	II	47.0	93	B	II	48.5
14	D	I	15.9	54	B	III	36.5	94	D	III	35.2
15	B	III	79.6	55	A	III	19.3	95	D	III	12.7
16	E	I	41.0	56	E	III	32.7	96	D	III	31.8
17	B	IV	81.4	57	A	II	13.6	97	E	II	21.0
18	D	II	60.7	58	E	I	10.9	98	B	IV	51.7
19	B	I	82.2	59	C	II	43.2	99	A	III	54.9
20	E	I	75.8	60	B	IV	54.4	100	A	III	21.6
21	D	I	47.3	61	D	I	32.8	101	B	III	26.5
22	E	III	70.9	62	E	IV	67.9	102	C	III	28.0
23	B	I	72.4	63	E	IV	28.2	103	E	IV	71.9
24	C	I	17.3	64	E	III	52.0	104	E	I	32.9
25	D	I	77.2	65	A	I	65.5	105	E	III	20.0
26	A	III	42.1	66	A	II	67.3	106	D	III	47.4
27	C	IV	75.4	67	C	III	61.7	107	B	III	49.3
28	C	II	36.6	68	D	II	59.7	108	C	II	38.0
29	E	I	38.1	69	B	II	59.5	109	A	III	43.6
30	B	II	56.2	70	D	I	71.1	110	A	II	18.1
31	C	I	59.7	71	A	I	11.4	111	D	II	38.1
32	A	II	60.6	72	C	II	40.3	112	D	I	48.8
33	D	IV	58.0	73	D	II	20.5	113	A	III	18.5
34	D	I	45.6	74	A	I	57.0	114	D	III	26.5
35	D	III	30.6	75	D	I	30.0	115	A	II	39.8
36	B	I	86.5	76	B	II	81.9	116	D	IV	46.6
37	A	II	58.0	77	E	III	56.1	117	B	III	35.0
38	B	III	50.1	78	C	II	60.6	118	D	III	36.9
39	C	II	63.2	79	D	II	73.9	119	B	III	24.5
40	B	I	51.6	80	D	III	73.3	120	A	II	28.0

Note: The percentages in Table 1 are based on the test records of 1,970 examinees who took the Mathematics Test at the November 1989 administration. The mean score for this group of examinees is MAT607 and the standard deviation of the scores is MAT83.

*See categories in Table 3, page 41.

Table 2: Mathematics Test
Conversion Table

Number Right	Scaled Score	Number Right	Scaled Score	Number Right	Scaled Score
0	320	40	510	80	700
1	330	41	520	81	710
2	330	42	520	82	710
3	330	43	520	83	710
4	340	44	530	84	720
5	340	45	530	85	720
6	350	46	540	86	730
7	350	47	540	87	730
8	360	48	550	88	740
9	360	49	550	89	740
10	370	50	560	90	750
11	370	51	560	91	750
12	380	52	570	92	760
13	380	53	570	93	760
14	390	54	580	94	770
15	390	55	580	95	770
16	400	56	590	96	780
17	400	57	590	97	780
18	410	58	600	98	790
19	410	59	600	99	790
20	420	60	610	100	800
21	420	61	610	101	800
22	430	62	610	102	800
23	430	63	620	103	810
24	430	64	620	104	810
25	440	65	630	105	820
26	440	66	630	106	820
27	450	67	640	107	830
28	450	68	640	108	830
29	460	69	650	109	840
30	460	70	650	110	840
31	470	71	660	111	850
32	470	72	660	112	850
33	480	73	670	113	860
34	480	74	670	114	860
35	490	75	680	115	870
36	490	76	680	116	870
37	500	77	690	117	880
38	500	78	690	118	880
39	510	79	700	119	890
				120	890

Estimating an Actual Score from Your Practice Score

When you actually take the Mathematics test, you will have questions to answer that will be similar to the questions in this book, but they will not be identical. Because of the differences in questions, the test that you actually take may be slightly more or less difficult than the test printed in this book. Therefore, you should not expect to get the same score that you have gotten on this test.

How to Interpret Your Score

The most meaningful interpretation of a test score occurs in the context of a particular use of the test. Thus, the interpretation of your score depends on the purpose for which you are taking the test. Scores may be interpreted in relation to some level of performance, such as a minimum score considered acceptable to satisfy established proficiency requirements. For example, if you are taking the Mathematics test because it is required for teacher certification in your state, the most useful information you could have would be the value of the qualifying score used by the teacher certifying agency in your state. Information about state qualifying scores or institutional requirements, which are determined by the state agencies and institutions, can be obtained directly from them.

Scores on the Mathematics test also may be interpreted in relation to the performance of appropriate reference groups of individuals who have taken the test. To assist you in interpreting your scaled score, percentile rank information based on data from the most recent three-year period of test administrations is provided in *Interpreting NTE Specialty Area Test Scores*, a leaflet that is sent with examinees' score reports. You can compare the score you obtained on the practice test in this book with those shown in the leaflet to see if you scored above or below the average. The leaflet is updated yearly and a revised version published each fall. It may be obtained by writing to NTE Programs, Educational Testing Service, P.O. Box 6051, Princeton, NJ 08541-6051.

Interpretive Information for Test Questions and Categories

The table listing the correct answers (Table 1) also lists the percentage of examinees choosing each correct answer. Thus, you can compare your performance on each question with that of the November 1989 group. The percentages give an indication of whether the group as a whole found a question easy (say, 75 percent chose the correct answer) or difficult (say, 25 percent chose the correct answer).

Information based on the percentages from Table 1 is summarized in Table 3, which lists the number of questions in each content category that fall within percentage ranges, the total number of questions in each category, the average (mean) number of questions answered correctly, and the average number of questions answered incorrectly. Thus, if you count the number of questions you answered correctly in each content category, you can compare how well you did with how the November 1989 group did in general on each category of the test.

Table 3: Content Categories of Mathematics Test

Frequency Distribution of Number of Questions in Each Category
Correctly Answered by Percentages of Examinees

	Percentage Ranges	Basic Math I	Pre-Calculus College Math II	Upper Division Math III	Professional Understanding and Pedagogy IV	Number of Questions Within each Range
"Easy" ↑	95+					
	90-94	2				2
	85-89	4				4
	80-84	1	1	2	1	5
	75-79	8			1	9
	70-74	5	3	2	1	11
	65-69	2	2	1	1	6
	60-64	1	7	1		9
	55-59	2	4	3	1	10
	50-54	2		2	2	6
	45-49	4	2	2	2	10
	40-44	1	3	3		7
	35-39	1	4	4		9
	30-34	3	2	3		8
	25-29	1	2	5	1	9
"Hard" ↓	20-24		2	4		6
	15-19	2	1	2		5
	10-14		1	1		4
	5-9					
Total Number of Questions		41	34	35	10	120
Mean Number Answered Correctly		24.4	16.2	14.1	5.7	60.4
Mean Number Answered Incorrectly*		15.8	16.2	18.8	3.9	54.7

* Questions that were not answered or were not attempted are not included as incorrect responses.

You can also do a similar comparison based on the number of questions you answered incorrectly. When you examine Table 3, note that some questions or categories may appear to be more difficult than others. The apparent difficulty of the questions (or category) may be affected not only by the actual difficulty of the questions themselves, but also by the ability level and training of the group on which the statistics are based.

Explanations of the Answers to the Practice Test Questions

1. $\dfrac{3x-1}{2x+1}$ is undefined if $x =$

   (A) $-\dfrac{1}{2}$

   (B) $-\dfrac{1}{3}$

   (C) 0

   (D) $\dfrac{1}{3}$

   (E) $\dfrac{1}{2}$

Since division by zero is undefined, an algebraic fraction is undefined if its denominator is zero. The denominator $2x+1=0$ when $x=-\dfrac{1}{2}$. The correct answer is A.

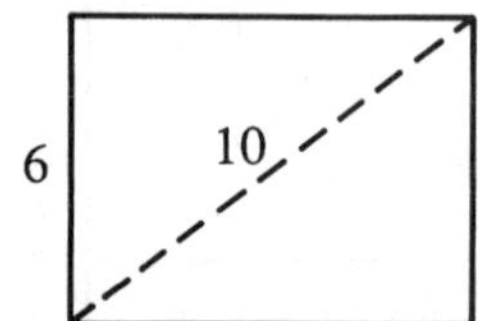

2. In the rectangle above, the perimeter is

   (A) 8
   (B) 24
   (C) 28
   (D) 38
   (E) 48

By the Pythagorean theorem, the length of the rectangle is $\sqrt{10^2-6^2}=8$; therefore, the perimeter is $2(6+8)=28$. The correct answer is C.

3. If 5 percent of x is 2.4, then 500 percent of x is

   (A) 12
   (B) 24
   (C) 120
   (D) 240
   (E) 2,400

If 5 percent of x is 2.4, then 500 percent of x equals $100(5\% \text{ of } x) = 100(2.4) = 240$. The correct answer is D.

4. If the average distance from the Earth to the Sun is 93,000,000 miles, and the average distance from the Earth to the Moon is 238,000 miles, then the Sun is about how many times as far from the Earth as the Moon is?

   (A)   400
   (B) 1,800
   (C) 4,000
   (D) 18,000
   (E) 40,000

The ratio of 93,000,000 to 238,000 is equal to the ratio of $\dfrac{93,000}{238}$, which is approximately $\dfrac{9,300}{24}$ or approximately 400. The best answer is A.

5. Which of the following fractions is equal to the infinite decimal $0.27\overline{27}$?

   (A) $\dfrac{3}{1,100}$

   (B) $\dfrac{1}{37}$

   (C) $\dfrac{3}{110}$

   (D) $\dfrac{27}{100}$

   (E) $\dfrac{3}{11}$

Let n represent the number $0.27\overline{27}$; then $100n = 27.27\overline{27}$ and $100n - n = 99n = 27$. Therefore, $n = \dfrac{27}{99} = \dfrac{3}{11}$. The correct answer is E.

6. Which of the following must be equal to zero for all real numbers x ?

   I. $-\dfrac{1}{x}$

   II. $x + (-x)$

   III. x^0

   (A) I only
   (B) II only
   (C) I and III only
   (D) II and III only
   (E) I, II, and III

Clearly, $-\dfrac{1}{x}$ can never be equal to zero since its numerator is not equal to zero. The expression $x+(-x)$ is equal to zero for all values of x; and x^0 is equal to 1 for all nonzero values of x. Therefore, only II can be, and must be, equal to zero for all real numbers x. The correct answer is B.

7. For $xy \neq 0$, $x^{-1} - y^{-2} =$

(A) $\dfrac{-1}{x} - \dfrac{-2}{y}$

(B) $\dfrac{1}{x - y^2}$

(C) $\dfrac{1}{x} - \dfrac{2}{y}$

(D) $\dfrac{y^2 - x}{xy^2}$

(E) $\dfrac{x - y^2}{xy^2}$

The expression $x^{-1} - y^{-2}$ is equivalent to $\dfrac{1}{x} - \dfrac{1}{y^2} = \dfrac{y^2 - x}{xy^2}$. The correct answer is D.

---

8. A bottling plant has an old machine that fills 6,000 bottles per hour and a new machine that fills 6,000 bottles every 30 minutes. How many minutes will it take both machines working together to fill 6,000 bottles?

(A) 20

(B) 30

(C) 45

(D) $66\dfrac{2}{3}$

(E) 90

The old machine fills $\dfrac{6,000}{60} = 100$ bottles per minute, and the new machine fills $\dfrac{6,000}{30} = 200$ bottles per minute. Working together, the two machines can fill $100 + 200 = 300$ bottles per minute. Therefore, it will take $\dfrac{6,000}{300} = 20$ minutes to fill 6,000 bottles. The correct answer is A.

9. If x, $x + 1$, and $x + 2$ are consecutive integers, which of the following must be true?

   I. The average of the three integers is divisible by 2.

   II. The sum of the three integers is divisible by the middle integer.

   III. $(x + 1)(x + 2) - x(x + 1)$ is divisible by 2.

(A) I only

(B) II only

(C) III only

(D) II and III only

(E) I, II, and III

In statement I, the average of the three integers is $\dfrac{x + (x + 1) + (x + 2)}{3}$ or $\dfrac{3x + 3}{3} = x + 1$, which is not necessarily divisible by 2. Therefore, statement I is not always true. In statement II, the sum of the three integers is $3x + 3$, which is a product of 3 and the middle integer, $x + 1$. Therefore, II must be true. In statement III, $(x + 1)(x + 2) - x(x + 1) = (x + 1)(x + 2 - x) = 2(x + 1)$, which is divisible by 2. Therefore, statements II and III must be true. The correct answer is D.

---

10. For $x \neq 0$, $\dfrac{5x - 3}{3x} + \dfrac{1}{x} =$

(A) $\dfrac{5x - 2}{4x}$

(B) $\dfrac{8x - 3}{3x}$

(C) $\dfrac{5x - 2}{3x}$

(D) $\dfrac{5}{3}$

(E) $-\dfrac{5}{3}$

For $x \neq 0$, $\dfrac{5x - 3}{3x} + \dfrac{1}{x} = \dfrac{5x - 3}{3x} + \dfrac{1}{x}\left(\dfrac{3}{3}\right) = \dfrac{5x - 3 + 3}{3x} = \dfrac{5}{3}$. The correct answer is D.

---

11. If the area of each face of a cube is 9, what is the volume of the cube?

(A) 18

(B) 27

(C) 54

(D) 81

(E) 729

If the area of a face of the cube is 9, then the edge of the cube is $\sqrt{9} = 3$ and the volume of the cube is $3^3 = 27$. The correct answer is B.

12. If x is a real number and $\sqrt{2x + 7} = x + 2$, then x could be

(A) -3 only
(B) -1 only
(C) 1 only
(D) -1 or -3
(E) -1 or 3

If x is a real number and $\sqrt{2x + 7} = x + 2$, then
$$2x + 7 = x^2 + 4x + 4$$
$$0 = x^2 + 2x - 3$$
$$0 = (x + 3)(x - 1)$$

Thus, $x = -3$ or $x = 1$. If $x = -3$, then $\sqrt{2(-3) + 7} = 1$, but $-3 + 2 = -1$. Therefore, $x = -3$ is an extraneous root that was introduced when both sides of the equation were squared, and only $x = 1$ satisfies the equation. The correct answer is C.

---

13. Of the following, which is closest to 196 ?

(A) $(8\sqrt{3})^2$
(B) $(6\sqrt{5})^2$
(C) $(5\sqrt{7})^2$
(D) $(9\sqrt{2})^2$
(E) $(5\sqrt{6})^2$

Calculating the value of each of the options in turn yields the following numbers: (A) $64(3) = 192$, (B) $36(5) = 180$, (C) $25(7) = 175$, (D) $81(2) = 162$, and (E) $25(6) = 150$. Since 192 is closest to 196, the correct answer is A.

14. A student divides as follows:

$$\frac{15}{32} \div \frac{3}{8} = \frac{15 \div 3}{32 \div 8} = \frac{5}{4}$$

Of the following, which is the best teacher response to the student?

(A) This is an incorrect method for division of fractions; the method works only in this specific problem.
(B) This is an incorrect method for division of fractions; neither of the fractions was inverted.
(C) This method will yield a correct answer only in special cases; the only method that will always yield a correct answer is to invert the second fraction and multiply.
(D) This method always yields an equivalent expression, but not always in its simplest form.
(E) This method will yield a correct answer only if the quotient is greater than 1.

To test whether this method will always yield a correct answer, consider the general case:

$$\frac{a}{b} \div \frac{c}{d} = \frac{ad}{bc} \quad \text{and} \quad \frac{a \div c}{b \div d} = \frac{\frac{a}{c}(dc)}{\frac{b}{d}(dc)} = \frac{ad}{bc}$$

Thus, this method yields an equivalent expression, or correct answer. However, the expression is not necessarily in its simplest form, since $a \div c$ and $b \div d$ need not be integers. The correct answer is D.

**FREQUENCY DISTRIBUTION OF SCORES
ON AN ALGEBRA TEST**

Test Score	Frequency
100	1
95	3
90	4
85	4
80	4
75	6
70	4
65	3
60	1
Total Number of Students	30

15. What is the mode of the distribution above?

    (A) 70
    (B) 75
    (C) 79
    (D) 80
    (E) 82

The mode of a distribution of scores is the score that occurs most frequently. Since the score that occurs most frequently in this distribution is 75, the correct answer is B.

---

16. Which of the following statements about measurement is (are) true?

    I. Both metric units and English units are standard units.
    II. Physical measurements are, at best, approximations.
    III. Physical measurements made with nonstandard units can be useful.

    (A) II only
    (B) I and II only
    (C) I and III only
    (D) II and III only
    (E) I, II, and III

Statement I is true because standards exist at the Bureau of Weights and Measures for both metric and English measurement units. Statement II is true because there is always some error in physical measurements, the amount depending on the precision of the instrument and the care taken by the person using the instrument. Statement III is true because, if the measurement need not be very precise, one may measure a distance in paces or by walking toe-to-heel, for example. The correct answer is E.

17. Which of the following sources would be most likely to have some suggestions on classroom-tested techniques for motivating students in a high school geometry class?

    (A) *Mathematical Reviews*
    (B) *Mathematics Teacher*
    (C) *The American Mathematical Monthly*
    (D) *Arithmetic Teacher*
    (E) *Transactions of the American Mathematical Society*

Both *Mathematics Teacher* and *Arithmetic Teacher* are journals published by the National Council of Teachers of Mathematics and have suggestions about techniques for motivating and teaching students in mathematics. *Arithmetic Teacher* focuses on elementary school mathematics, and *Mathematics Teacher* focuses on secondary school mathematics. The other three journals listed are of greatest interest to university-level mathematics teachers and to nonteaching mathematicians. Therefore, *Mathematics Teacher* would be most likely to provide helpful suggestions for a high school geometry class. The best answer is B.

---

18. If $\log_x 25 = \frac{2}{3}$, then $x =$

    (A) -5
    (B) 5
    (C) 25
    (D) 125
    (E) 625

If $\log_x 25 = \frac{2}{3}$, then $x^{\frac{2}{3}} = 25$.

Thus, $x = 25^{\frac{3}{2}} = \left(\sqrt{25}\right)^3 = 5^3 = 125$. The correct answer is D.

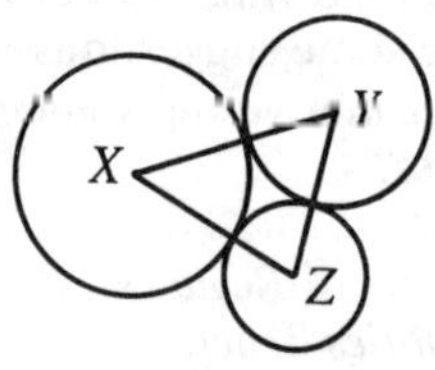

19. In the figure above, X, Y, and Z are the centers of circles that have diameters of 24, 14, and 10, respectively. What are the lengths of the sides of $\triangle XYZ$?

(A) 12, 7, and 5
(B) 19, 17, and 12
(C) 24, 14, and 10
(D) 38, 34, and 24
(E) 48, 28, and 20

If the circles with centers X, Y, and Z have diameters 24, 14, and 10, respectively, they have radii 12, 7, and 5, respectively. Thus, side XY has length $12+7=19$, side XZ has length $12+5=17$, and side YZ has length $7+5=12$. Therefore, the correct answer is B.

———————————

20. If $2xy = 2x$, which of the following must be true?

(A) $y = 0$
(B) $x = 0$
(C) $x = y = 0$
(D) $x = y = 1$
(E) $x = 0$ or $y = 1$

If $2xy = 2x$, then $2xy - 2x = 0$ and so $2x(y-1) = 0$. This equation implies that either $x = 0$ or $y = 1$. The correct answer is E.

———————————

21. $2^9 \cdot 5^7$ is an integral multiple of all of the following EXCEPT

(A)   1
(B)   2
(C)   8
(D)   75
(E)   100

If $\left(2^9\right)\left(5^7\right)$ is an integral multiple of a number n, then n is a factor of $\left(2^9\right)\left(5^7\right)$. Clearly, 1, 2, $2^3 = 8$, and $\left(2^2\right)\left(5^2\right) = 100$ are all factors of $\left(2^9\right)\left(5^7\right)$. However, $(3)\left(5^2\right) = 75$ is not a factor of $\left(2^9\right)\left(5^7\right)$ since $\left(2^9\right)\left(5^7\right)$ does not have the factor 3. The correct answer is D.

22. Which of the following is FALSE for all real numbers?

(A) If $x + y > z$, then $x > z - y$.

(B) If $\frac{x}{y} > z$ and $y > 0$, then $x > yz$.

(C) If $x - y > 0$, then $x > y$.

(D) If $x < y$ and $x > 0$, then $x^2 < y^2$.

(E) If $-3x < 6$, then $x < -2$.

Statements A and C are true, since adding $-y$ or y to each side of an inequality produces an equivalent inequality. Statement B is true, since multiplying each side of an inequality by a positive number produces an equivalent inequality. Statement D is true, since both x and y are positive and squaring each side of such an inequality produces an equivalent inequality. Statement E is false, since multiplying or dividing each side of an inequality by a negative number reverses the order of the inequality. The best answer is E.

———————————

23. Of the following numbers, which is the arithmetic mean of 4 consecutive multiples of 3 ?

(A)   5

(B) $7\frac{1}{2}$

(C)   9

(D) $10\frac{1}{3}$

(E)   12

Let the 4 consecutive multiples of 3 be $3k$, $3(k+1)$, $3(k+2)$, and $3(k+3)$. Then the sum of these numbers is $3k + (3k+3) + (3k+6) + (3k+9) = 12k + 18$, and the arithmetic mean is $\frac{12k+18}{4} = 3k + \frac{18}{4} = 3k + 4\frac{1}{2}$. Inspection of the options shows that only $7\frac{1}{2}$ can be written in the form $3k + 4\frac{1}{2}$, where $k = 1$. Therefore, the correct answer is B.

24. To determine whether a positive integer N is a prime number, it is necessary and sufficient to check whether N is divisible by any prime numbers less than or equal to

(A) N

(B) N^2

(C) $\sqrt{N}$

(D) $\frac{1}{2}N$

(E) $N - 1$

Note that if d is a proper factor of N that is greater than $\sqrt{N}$, then $\dfrac{N}{d}$ is a proper factor less than $\sqrt{N}$. Thus, N is a prime number if, and only if, it has no prime factor less than or equal to $\sqrt{N}$. The correct answer is C.

---

25. If a student states that

$$-\frac{1}{2}\left(-\frac{2}{3}x + \frac{1}{2}\right) = \frac{1}{3}x + \frac{1}{2},$$

it is most likely that the mistake results from a misunderstanding of which of the following?

(A) Multiplication of fractions
(B) Arithmetic of negative numbers
(C) Commutative property of multiplication
(D) Distributive property
(E) Equivalence relation

Since $\left(-\dfrac{1}{2}\right)\left(-\dfrac{2}{3}x\right) = \dfrac{1}{3}x$, the student has demonstrated the understanding of multiplication of fractions and the arithmetic of negative numbers. Therefore, the answer cannot be A or B; choices C and E are not involved in the example, so no conclusions can be drawn about these.

However, since the student did not multiply the second term of the sum by $-\dfrac{1}{2}$, but apparently thinks that

$-\dfrac{1}{2}\left(-\dfrac{2}{3}x + \dfrac{1}{2}\right) = \left(-\dfrac{1}{2}\right)\left(-\dfrac{2}{3}x\right) + \dfrac{1}{2}$, it is most likely that the student misunderstands the distributive property. The best answer is D.

26. If $A = \begin{pmatrix} 1 & 0 \\ 1 & 1 \end{pmatrix}$ and $B = \begin{pmatrix} 1 & 1 \\ 0 & 1 \end{pmatrix}$, then $AB =$

(A) $\begin{pmatrix} 1 & 1 \\ 1 & 2 \end{pmatrix}$

(B) $\begin{pmatrix} 2 & 1 \\ 1 & 1 \end{pmatrix}$

(C) $\begin{pmatrix} 2 & 1 \\ 1 & 2 \end{pmatrix}$

(D) $\begin{pmatrix} 1 & 0 \\ 0 & 1 \end{pmatrix}$

(E) $\begin{pmatrix} 1 & 1 \\ 1 & 1 \end{pmatrix}$

For any two n by n matrices A and B, the ij entry of the product AB is obtained by taking the inner (dot) product of the ith row of A with the jth column of B.

Thus, the product $\begin{pmatrix} 1 & 0 \\ 1 & 1 \end{pmatrix}\begin{pmatrix} 1 & 1 \\ 0 & 1 \end{pmatrix} =$

$\begin{pmatrix} 1(1)+0(0) & 1(1)+0(1) \\ 1(1)+1(0) & 1(1)+1(1) \end{pmatrix} = \begin{pmatrix} 1 & 1 \\ 1 & 2 \end{pmatrix}$.

The correct answer is A.

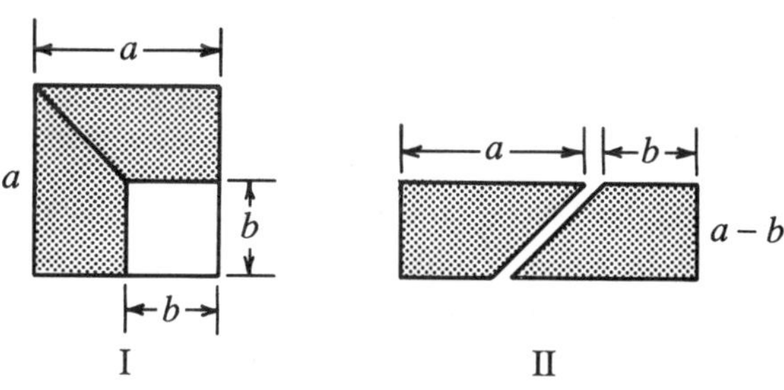

27. In the figures above, the two shaded trapezoidal regions of the square in Figure I can be rearranged as shown in Figure II. This illustration can be used to show students a geometric verification of which of the following identities?

(A) $a^2 + ab = a(a + b)$

(B) $a^2 - ab = a(a - b)$

(C) $a^2 - b^2 = (a + b)(a - b)$

(D) $a^2 - 2ab + b^2 = (a - b)^2$

(E) $a^2 + 2ab + b^2 = (a + b)^2$

The area of the shaded region in Figure I is equal to the difference in the areas of the larger square (a^2) and the smaller square (b^2). When one of the trapezoidal regions is flipped and the two regions are rearranged as in Figure II, the parts form a rectangular region that has area $(a + b)(a - b)$. Thus, this illustration is a geometric interpretation of $a^2 - b^2 = (a + b)(a - b)$. Of the identities listed as options, the only one that is relevant to the illustration is C.

$$\begin{cases} x^2 + y^2 = 25 \\ x + y = 0 \end{cases}$$

28. The pair of equations above has how many common solutions?

(A) None
(B) Exactly 1
(C) Exactly 2
(D) Exactly 4
(E) Infinitely many

The graph of $x^2 + y^2 = 25$ is a circle of radius 5 with center (0,0); the graph of $x + y = 0$ is a line with slope -1 that passes through (0,0). Since the number of common solutions is the number of points where the graphs of the circle and the line intersect, the correct answer is C.

---

29. When the base of a triangle is decreased by 20 percent and the altitude to this base is increased by 20 percent, the area is

(A) increased by 4%
(B) increased by 2%
(C) unchanged
(D) decreased by 2%
(E) decreased by 4%

If b represents the base and h represents the height of the original triangle, then $\frac{1}{2}bh$ represents its area. When the base is decreased by 20 percent and the altitude is increased by 20 percent, the area of the new triangle is $\frac{1}{2}(0.8b)(1.2h) = 0.96\left(\frac{1}{2}bh\right)$, which is 96 percent of the area of the original triangle. Thus, the area of the new triangle is $100\% - 96\% = 4\%$ less than the area of the original triangle. The correct answer is E.

30. If $\tan(t + k) = \tan t$ for all real numbers t for which $\tan(t + k)$ and $\tan t$ exist, then k could be

(A) $\dfrac{3\pi}{2}$

(B) π

(C) $\dfrac{2\pi}{3}$

(D) $\dfrac{\pi}{2}$

(E) $\dfrac{\pi}{4}$

Because the period of the function $y = \tan t$ is π, $\tan t = \tan(t + k)$ for all $k = n\pi$, where n is an integer and the functions $\tan t$ and $\tan(t + k)$ are defined. (See the graph below.) Therefore, of the options given, k could only be equal to π, and the correct answer is B.

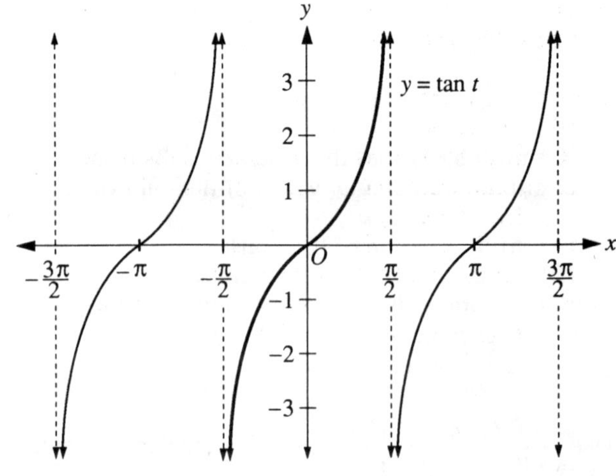

---

31. If the area of an isosceles right triangle is 32, what is the perimeter of the triangle?

(A) $12\sqrt{2}$
(B) $8 + 8\sqrt{2}$
(C) $16 + 8\sqrt{2}$
(D) 32
(E) 64

Because the legs of a right isosceles triangle have the same length, the area of such a triangle is $\frac{1}{2}x^2$, where x is the length of each leg. If $\frac{1}{2}x^2 = 32$, then $x^2 = 64$ and $x = 8$. If the length of each leg is 8, then the length of the hypotenuse is $\sqrt{8^2 + 8^2} = \sqrt{128} = 8\sqrt{2}$. Therefore, the perimeter of the triangle is $8 + 8 + 8\sqrt{2} = 16 + 8\sqrt{2}$, and the correct answer is C.

32. If $r > s > 1$, then $\dfrac{\log r - \log s}{\log\left(\dfrac{r}{s}\right)} - 1$ equals

    (A) 0

    (B) $\dfrac{1}{2}$

    (C) 1

    (D) $\dfrac{r}{s(r - s)}$

    (E) $\dfrac{s(r - s)}{r}$

If $rs \neq 0$, then $\log r - \log s = \log\left(\dfrac{r}{s}\right)$. Thus, the expression to be evaluated is equivalent to $1 - 1 = 0$, and the correct answer is A.

---

33. Of the following mathematicians, which one is credited with the creation of coordinate geometry?

    (A) Leonhard Euler
    (B) Leonardo Fibonacci
    (C) Evariste Galois
    (D) René Descartes
    (E) Karl Gauss

The rectangular coordinate system is often referred to as the Cartesian coordinate system in honor of René Descartes, who is credited with its creation. The correct answer is D.

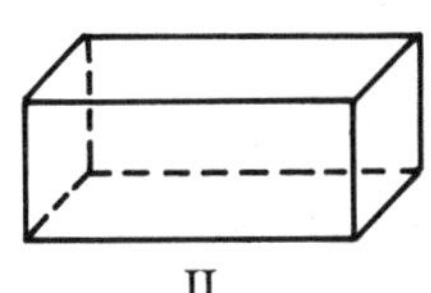

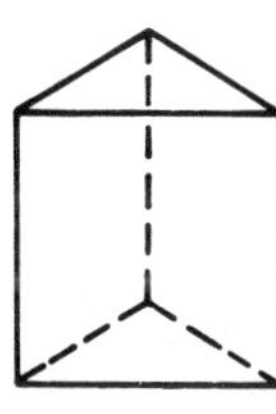
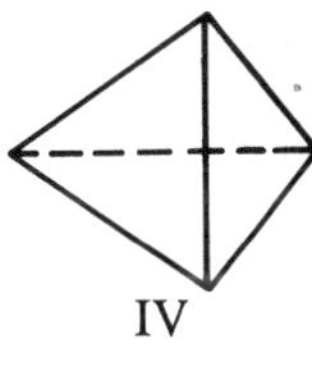

34. The volume(s) of which of the solids shown above can be found by applying the formula $V = Bh$, where V represents the volume, B represents the area of a base, and h represents the altitude to the base that has area B?

    (A) II only
    (B) I and II only
    (C) I and III only
    (D) I, II, and III only
    (E) I, II, III, and IV

The volume of a solid can be found by multiplying the area of its base, B, by its height, h, provided that (1) the solid has at least two congruent faces that lie in parallel planes, and (2) every cross section that is parallel to the planes of the congruent faces will be congruent to these faces. All prisms and cylinders have these two properties. Thus, the figures given that have these properties are I, II, and III. The correct answer is D.

---

35. Two answer sheets are different if at least one answer on the two sheets is different. How many different answer sheets are possible for a test consisting of 10 questions, all of which are answered true or false?

    (A)   20
    (B)   64
    (C) 100
    (D) 1,024
    (E) More than 10,000

Each of the 10 questions can be answered in either of two ways, true or false. Therefore, the total number of different answer sheets possible is $2^{10} = 1,024$. The correct answer is D.

36. If $40{,}404 - (4{,}040{,}400 \times 10^k) = 0$, then $k =$

(A) -3
(B) -2
(C) -1
(D) 2
(E) 4

Since $40{,}404 = 4{,}040{,}400 \times 10^k$, it follows that $\dfrac{40{,}404}{4{,}040{,}400} =$

$10^k = \dfrac{1}{100} = 10^{-2}$. Thus, $k = -2$, and the correct answer is B.

---

37. Which of the following represents the graph of
$$y = \frac{|x + 1|}{x + 1}?$$

(A) 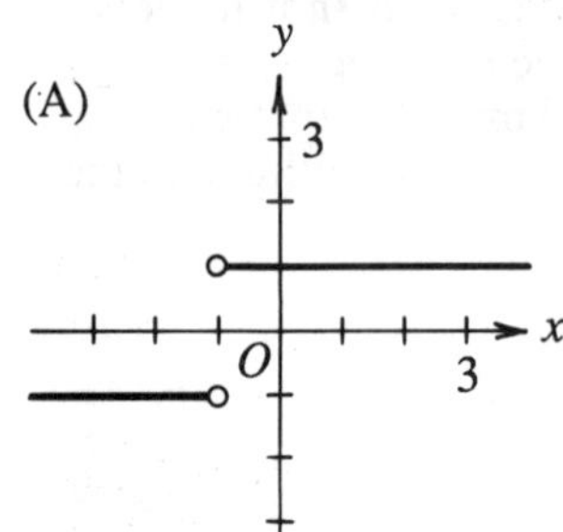

(B) 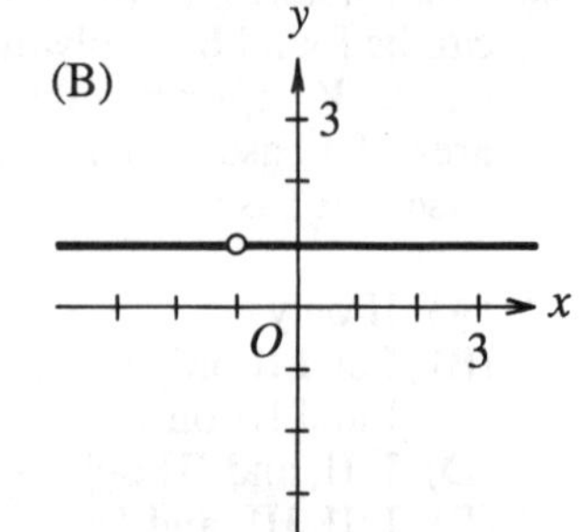

(C) 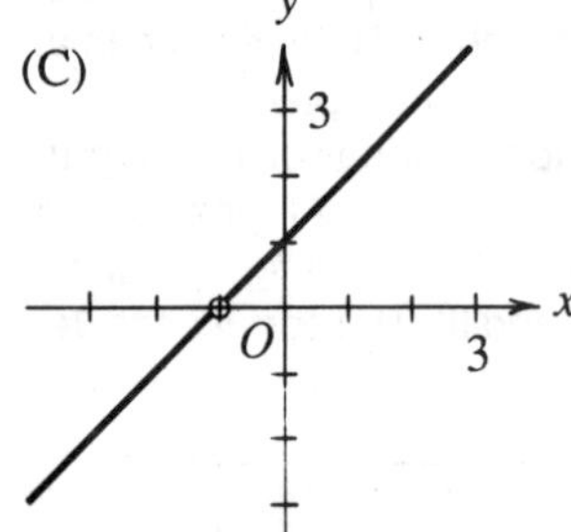

(D) 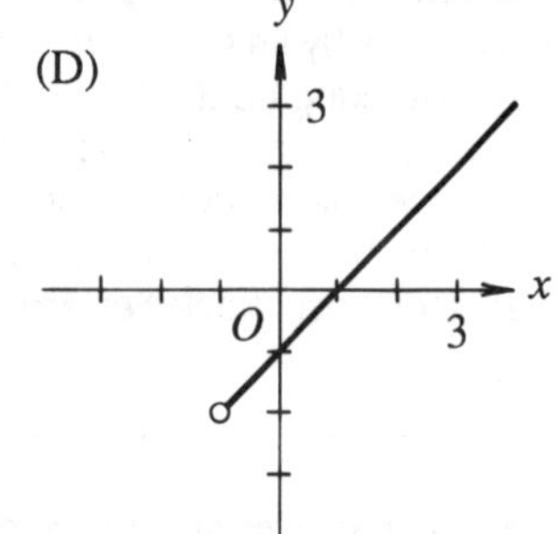

(E) 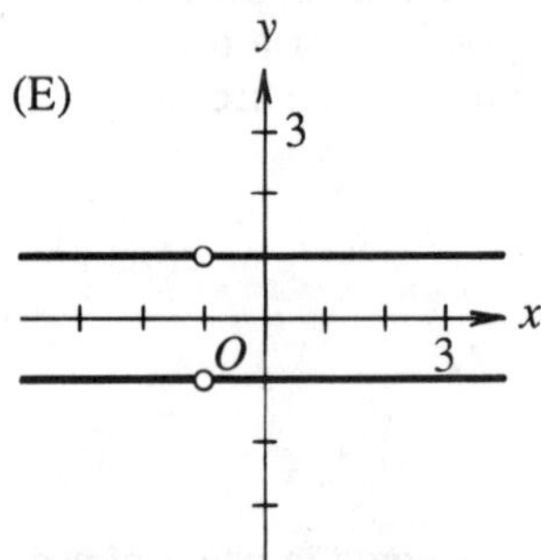

For all $x > -1$, $|x + 1| = x + 1$, so that $y = 1$. For $x < -1$, $|x + 1| = -(x + 1)$, so that $y = -1$. When $x = -1$, the function is undefined. The graph that has these characteristics is shown in option A.

38. Which of the following methods would be appropriate in proving the conjecture, "All positive integers have property S"?

   I. Finding a nonpositive integer that has property S
   II. Showing that S is false for any nonpositive integer
   III. Using the principle of mathematical induction

(A) II only
(B) III only
(C) I and II only
(D) I and III only
(E) I, II, and III

Since the conjecture that all positive integers have property S is not equivalent to any conjecture about the negative integers and property S, methods I and II are not appropriate methods in proving the conjecture. Method III, using the principle of mathematical induction, would be an appropriate way to prove the conjecture; for example, one could verify that 1 (the first positive integer) has property S, assume that n has property S, and then prove that this implies that $(n + 1)$ has property S. Therefore, the best answer is B.

---

39. The equation of the line that is parallel to the line $x + 3 = y - 2$ and contains the point $(2, -3)$ is

(A) $x + y + 1 = 0$
(B) $x - y - 1 = 0$
(C) $x - y - 5 = 0$
(D) $x + y - 5 = 0$
(E) $x + y + 5 = 0$

The slope-intercept form of the equation $x + 3 = y - 2$ is $y = x + 5$. Since the slope of this line (the coefficient of x in the slope-intercept form) is 1, any line parallel to this line must have slope 1. Therefore, the equation of the line that we are looking for is of the form $y = x + b$. If the line contains the point $(2, -3)$, then $-3 = 2 + b$ and $b = -5$. Thus, the equation of the line with slope 1 and containing the point $(2, -3)$ is $y = x - 5$ or $x - y - 5 = 0$. The correct answer is C.

---

40. If x and y are integers and $x + y = 56$, which of the following must be true of $x - y$?

(A) $x - y$ is odd.
(B) $x - y$ is even.
(C) $x - y < 0$
(D) $x - y > 0$
(E) $x - y < 56$

Because the sum of x and y is an even number, x and y must have the same parity (i.e., either both are even or both are odd). In either case, $x - y$ must be an even number, and the correct answer is B. Any of the choices C, D, and E could be true, but none *must* be true, since no information is given about the values of x and y individually.

41. A student concludes that since $\frac{3}{3} = 1$ and $\frac{5}{5} = 1$, then $\frac{0}{0} = 1$. Which of the following would be the most appropriate way for the teacher to respond to this student's conclusion?

(A) To agree and explain that a number divided by itself is always equal to one
(B) To agree and show that checking the division by multiplication confirms that $0 \cdot 1 = 0$
(C) To disagree and explain that zero divided by any number is always equal to zero
(D) To disagree and explain that a unique result cannot be found since $0 \cdot a = 0$ for any value of a
(E) To disagree and explain that any number divided by zero is infinite

Since division by zero is undefined, to agree with the student's conclusion as in choices A and B would clearly be wrong. The explanation given in choice C ignores the fact that division by zero is undefined and focuses instead on the zero in the numerator. The explanation given in choice D should help the student to understand why division by zero is undefined. The explanation given in choice E confuses $\lim\limits_{x \to 0+} \frac{N}{x} = \infty$ with the division of a number N by zero itself. Choice D is clearly the best answer.

---

42. The initial point of the vector $\overrightarrow{PT}$ is $P(1, 3)$ and the terminal point is $T(4, -2)$. What is the terminal point of a vector that is equivalent to $\overrightarrow{PT}$ and has the origin as its initial point?

(A) $(-3, 5)$
(B) $(1, -3)$
(C) $(3, -5)$
(D) $(5, 1)$
(E) $(5, 5)$

Two vectors are equivalent if they have the same length and direction. To go from the initial point to the terminal point of $\overrightarrow{PT}$, the x-coordinate of P is increased by 3 and the y-coordinate of P is decreased by 5. An equivalent vector with initial point $(0,0)$ will thus have terminal point $(3, -5)$. Therefore, the correct answer is C. See the figure below for the two vectors.

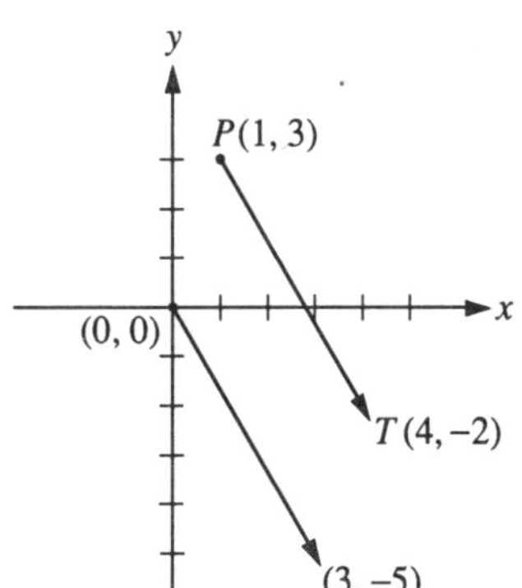

43. Which of the following sets of numbers is closed under the given operation?

(A) The real numbers under division
(B) The irrational numbers under addition
(C) The irrational numbers under multiplication
(D) The odd integers under addition
(E) The odd integers under multiplication

A set S is closed under a given operation, *, if and only if for all $a, b \in S$, $a * b \in S$. A proof is required to show that a particular set is closed under a particular operation; however, a single counterexample can be used to demonstrate that a set is not closed under a particular operation. Note the following counterexamples for choices A through D and the proof of closure for choice E.

(A) 4 and 0 are real numbers, but $\frac{4}{0}$ is not a real number.

(B) $(1+\sqrt{2})$ and $(1-\sqrt{2})$ are irrational numbers, but $(1+\sqrt{2})+(1-\sqrt{2})= 2$ is not an irrational number.

(C) $\sqrt{2}$ is an irrational number, but $\sqrt{2} \times \sqrt{2} = 2$ is not an irrational number.

(D) 1 and 3 are odd integers, but $1 + 3 = 4$ is not an odd integer.

(E) An odd integer is of the form $2n + 1$, where n is an integer. The product of two odd integers may be written as $(2a + 1)(2b + 1)$, where a and b are integers. Then $(2a + 1)(2b + 1) = 4ab + 2a + 2b + 1 = 2(2ab + a + b) + 1$, which is of the form $2n + 1$ and is, therefore, an odd integer. The correct answer is E.

---

44. If $\displaystyle\sum_{i=1}^{n} i = \frac{n(n + 1)}{2}$, then $\displaystyle\sum_{i=1}^{100} 3i =$

(A) $\dfrac{(100)(101)}{2} + 300$

(B) 100^3

(C) $\dfrac{(100)(101)(102)}{6}$

(D) $\dfrac{(300)(301)}{2}$

(E) $3\left[\dfrac{(100)(101)}{2}\right]$

Since $\displaystyle\sum_{i=1}^{n} i = \frac{n(n+1)}{2}$, it follows that $\displaystyle\sum_{i=1}^{n} 3i = 3\sum_{i=1}^{n} i = \frac{3n(n+1)}{2}$.

Substituting $n = 100$ into this formula yields

$\displaystyle\sum_{i=1}^{100} 3i = 3\left[\frac{(100)(101)}{2}\right]$. Therefore, the correct answer is E.

45. The product of complex numbers
$(2 + 3i)(3 - 2i)$ equals

(A)   0
(B)   12
(C) $5i$
(D) $6 - 6i$
(E) $12 + 5i$

$$(2+3i)(3-2i) = 6 - 4i + 9i - 6i^2$$
$$= 6 + (-4+9)i - 6(-1)$$
$$= 12 + 5i$$

Thus, the correct answer is E.

___

46. $\dfrac{6!}{3!} =$

(A) 5!
(B) 4!
(C) 3(3!)
(D) 3
(E) 2

$$\frac{6!}{3!} = \frac{6 \cdot 5 \cdot 4 \cdot 3 \cdot 2 \cdot 1}{3 \cdot 2 \cdot 1} = 5 \cdot 4 \cdot 3 \cdot 2 \cdot 1 = 5!$$

Thus, the correct answer is A.

___

47. When a student solves the system of equations $x = 3y$ and $3y = 2x - 1$ by first writing $x = 2x - 1$, the student is using which of the following properties?

(A) Reflexive
(B) Symmetric
(C) Transitive
(D) Commutative
(E) Associative

The student has taken two quantities, each of which is equal to $3y$, and set them equal to each other. This is an application of the transitive property, "If $a = b$ and $b = c$, then $a = c$." The correct answer is C.

___

48. If one assumes that a student A years old requires H hours of sleep per day, and that $H = 14 - \dfrac{A}{3}$, then how old is a student who requires 9 hours of sleep per day?

(A) 18
(B) 15
(C) 11
(D) 9
(E) 8

Substituting 9 for H in the formula $H = 14 - \dfrac{A}{3}$ yields

$9 = 14 - \dfrac{A}{3}$, or $A = 3(14 - 9) = 15$, and the correct answer is B.

49. What are all values of x such that $x < \sqrt{x}$?

(A) $0 < x < 1$
(B) $0 < x$
(C) $1 < x$
(D) $x < 0$
(E) $-1 < x < 0$

Since the inequality involves $\sqrt{x}$ and it is a strict inequality, it follows that $x > 0$.

Now $x < \sqrt{x} \Rightarrow (x)^2 < (\sqrt{x})^2 \Rightarrow x^2 < x \Rightarrow \dfrac{x^2}{x} < \dfrac{x}{x}$ or $x < 1$.

Combining the two inequalities yields $0 < x < 1$. Therefore, the correct answer is A.

___

50. The graph of which of the following equations is a line that is perpendicular to the line represented by the equation $3x + 2y = 5$?

(A) $y = -\dfrac{3}{2}x - 5$

(B) $y = -\dfrac{2}{3}x + 7$

(C) $y = -\dfrac{5}{2}x + \dfrac{3}{2}$

(D) $y = \dfrac{3}{2}x + 7$

(E) $y = \dfrac{2}{3}x + 3$

The slope of any line perpendicular to a nonhorizontal line with slope m is $-\dfrac{1}{m}$. Putting the equation $3x + 2y = 5$ into slope-intercept form yields the equation $y = -\dfrac{3}{2}x + \dfrac{5}{2}$. The slope of the given line is $-\dfrac{3}{2}$, and the slope of any line perpendicular to the given line is $-\left[-\dfrac{2}{3}\right]$, or $\dfrac{2}{3}$. Since all the equations in the options are in slope-intercept form, it is clear that only E gives an equation of a line that has slope $\dfrac{2}{3}$.

51. If $f(x) = \sqrt{x^2 + 7}$ for all real numbers x,
then $f'(3) =$

(A) $\dfrac{1}{8}$

(B) $\dfrac{1}{4}$

(C) $\dfrac{3}{4}$

(D) 2

(E) 12

If $f(x) = (x^2 + 7)^{\frac{1}{2}}$, then using the chain rule to calculate the first derivative yields $f'(x) = \dfrac{1}{2}(x^2 + 7)^{-\frac{1}{2}}(2x) = \dfrac{x}{\sqrt{x^2 + 7}}$. Thus, $f'(3) = \dfrac{3}{\sqrt{9+7}} = \dfrac{3}{\sqrt{16}} = \dfrac{3}{4}$, and the correct answer is C.

____

52. Circle K and square S each have area π. What is the ratio of the circumference of K to the perimeter of S?

(A) $\dfrac{\sqrt{\pi}}{4}$

(B) $\dfrac{2}{\pi}$

(C) $\dfrac{\pi}{4}$

(D) $\dfrac{\sqrt{\pi}}{2}$

(E) $\dfrac{\pi^2}{16}$

If circle K has area π, then it has radius 1 and circumference 2π. If square S has area π, then its sides have length $\sqrt{\pi}$ and its perimeter is $4\sqrt{\pi}$. Then the ratio of the circumference of K to the perimeter of S is $\dfrac{2\pi}{4\sqrt{\pi}} = \dfrac{\sqrt{\pi}}{2}$. Thus, the correct answer is D.

53. Which of the following is NOT a one-to-one function on the set of all real values of x for which the expression is defined?

(A) $f(x) = |x|$

(B) $f(x) = \sqrt{x}$

(C) $f(x) = 2^x$

(D) $f(x) = 2x + 1$

(E) $f(x) = x^3$

A function f such that $f(x_1) \neq f(x_2)$ whenever $x_1 \neq x_2$ is a one-to-one function. $f(x) = |x|$ is a function for which $f(x) = f(-x)$, and clearly $x \neq -x$ unless $x = 0$. Thus, this function is not one-to-one. Therefore, the correct answer is A.

Alternately, a function f is one-to-one if any line drawn parallel to the x-axis intersects the graph of the function in at most one point. If one examines the graphs of the five functions, only $f(x) = |x|$ does not satisfy this condition.

____

54. Which of the following sets of numbers has a greatest element?

(A) The integers
(B) The negative integers
(C) The rational numbers
(D) The negative rational numbers
(E) The negative real numbers

The greatest negative integer is −1. None of the other sets has a greatest element. Therefore, the correct answer is B.

o	a	b	c	d
a	d	c		
b	c			
c			c	
d	b			

55. The table above shows some of the entries in the multiplication table for a group $G = \{a, b, c, d\}$, with multiplication denoted by o. What is the multiplicative inverse of b?

(A) a
(B) b
(C) c
(D) d
(E) It cannot be determined from the information given.

To determine which of the elements of G is the multiplicative inverse of b, it is necessary first to determine the identity element of G, since any element times its inverse must equal the identity element. The identity element, e, of a group satisfies the equations $e \circ g = g \circ e = g$ for every element, g, of the group. Since $a \circ a$, $b \circ a$, and $d \circ a$ are all not equal to a, neither a, nor b, nor d is the identity element of G. Since every group has an identity element, c must be the identity element of G. (Alternately, since every element of a group has an inverse and since $c^2 = c$, it follows that $c^{-1}(c^2) = c^{-1}(c)$. Thus, c is the identity element of G.) It is now left to determine which element of the group multiplied (on either side) by b equals c, the identity element. Looking at the table shows $b \circ a = c$, and the best answer is A.

56. At $x = \frac{1}{3}$, the graph of $f(x) = x^3 - x^2 - x - 1$ has which of the following?

(A) A relative minimum
(B) A relative maximum
(C) An asymptote
(D) An x-intercept
(E) A point of inflection

It is efficient to analyze the graphic properties of this function by first considering the properties determinable from the equation of the function itself, then from the equations of its first and second derivatives. The graph of $f(x) = x^3 - x^2 - x - 1$ is the graph of a cubic polynomial. Since cubic polynomials have no asymptotes, C is not the correct answer.

If $\left(\frac{1}{3}, 0\right)$ is an x-intercept, then $f\left(\frac{1}{3}\right) = 0$. But

$$f\left(\frac{1}{3}\right) = \left(\frac{1}{3}\right)^3 - \left(\frac{1}{3}\right)^2 - \frac{1}{3} - 1 = \frac{1}{27} - \frac{1}{9} - \frac{1}{3} - 1 < 0, \text{ so D is not}$$

the correct answer.

If $\left(\frac{1}{3}, 0\right)$ is a relative maxima or minima, then $f'\left(\frac{1}{3}\right) = 0$.

But $f'(x) = 3x^2 - 2x - 1$ and

$$f'\left(\frac{1}{3}\right) = 3\left(\frac{1}{3}\right)^2 - 2\left(\frac{1}{3}\right) - 1 = \frac{1}{3} - \frac{2}{3} - 1 = -\frac{4}{3} \neq 0, \text{ so neither A}$$

nor B is the correct answer.

If $\left(\frac{1}{3}, 0\right)$ is a point of inflection, then $f''\left(\frac{1}{3}\right) = 0$. Taking the

second derivative of $f(x)$ yields $f''(x) = 6x - 2$. Thus,

$$f''\left(\frac{1}{3}\right) = 6\left(\frac{1}{3}\right) - 2 = 0, \text{ and the correct answer is E.}$$

---

57. T is the composite transformation of the coordinate plane that consists of a reflection in the x-axis followed by a reflection in the line through $(0, 0)$ and $(1, 1)$. If (x, y) is an arbitrary point in the plane, what is its image under T?

(A) $(-y, x)$
(B) $(-y, -x)$
(C) $(y, -x)$
(D) $(x, -y)$
(E) (x, y)

A reflection in the x-axis maps (x, y) to $(x, -y)$. A reflection in the line $y = x$ maps $(x, -y)$ to $(-y, x)$. Therefore, the correct answer is A.

58. If $N = \sqrt{(x-2)^2}$, which of the following must be true?

    I. $N = x - 2$
    II. $N = |x - 2|$
    III. $N \geq x - 2$

(A) I only
(B) II only
(C) III only
(D) I and III
(E) II and III

By definition, $\sqrt{N^2} = |N|$; therefore, $\sqrt{(x-2)^2}$ equals $|x-2|$, which is nonnegative. Thus, $N \neq x - 2$ if $x - 2 < 0$. Consequently, II must be true, but I need not be true. With respect to III, if $x - 2$ is negative, $N = |x - 2| > x - 2$, and if $x - 2$ is nonnegative, $N = x - 2$. Thus, III must be true, and the correct answer is E.

59. The point P can be written in polar coordinates as $\left(2, \dfrac{3\pi}{4}\right)$. How is point P written in rectangular coordinates?

(A) $(-2, -2)$

(B) $(-2, 2)$

(C) $(-\sqrt{2}, \sqrt{2})$

(D) $(-\sqrt{2}, -\sqrt{2})$

(E) $(\sqrt{2}, -\sqrt{2})$

To determine the rectangular coordinates (x, y) of a point $P(r, \theta)$ given in polar coordinates, recall that $x = r\cos\theta$ and $y = r\sin\theta$. Thus, in this case, since $\dfrac{3\pi}{4}$ radians equals $135°$, $x = r\cos\theta = 2\cos 135° = -2\cos 45° = -2 \cdot \dfrac{\sqrt{2}}{2} = -\sqrt{2}$ and $y = r\sin\theta = 2\sin 135° = 2\sin 45° = 2 \cdot \dfrac{\sqrt{2}}{2} = \sqrt{2}$. Therefore, the best answer is C.

Alternately, if a point P has polar coordinates (r, θ), then P is on a circular arc that has the origin O as its center and has radius r; angle θ represents the radian measure of the arc from the positive ray of the x-axis to point P. Thus, the point $(2, \dfrac{3\pi}{4})$ is a point that is 2 units from the origin and on a circular arc of $\dfrac{3\pi}{4}$ radians or $\dfrac{3}{4}(180°) = 135°$, as shown in the figure below. Note that in the figure $\angle POM = 45°$, which implies that $OM = PM$. Thus, if $OM = PM = k$, then $k = \sqrt{2}$, and the rectangular coordinates of P are $(-\sqrt{2}, \sqrt{2})$.

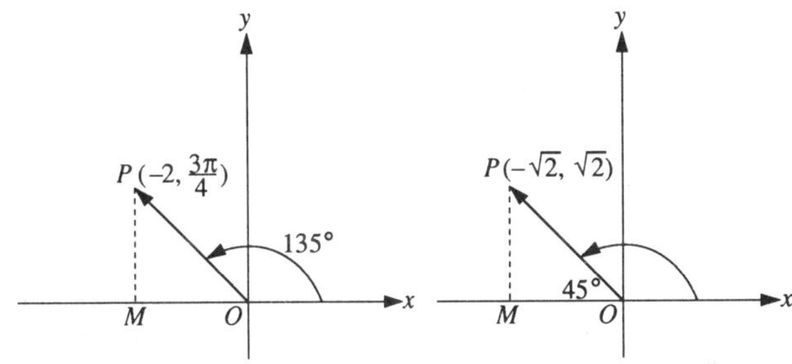

P in polar coordinates   P in rectangular coordinates

60. Of the following, which is most relevant to a classroom discussion of conic sections?

   (A) The law of refraction
   (B) The orbit of a planet
   (C) Ohm's law
   (D) Compound interest
   (E) Enlargement of a photograph

An ellipse is a conic section, and it is known that the orbit of a planet is roughly elliptical. Because B is the only choice that relates to conic sections, it is the best answer.

---

61. If (m, n) represents the greatest common divisor of the positive integers m and n, which of the following statements is true if $m \neq n$?

   (A) $(m, n) > 1$
   (B) $(m, n) > m$
   (C) $(m, n) < n$
   (D) $(m, n) < mn$
   (E) None of the above

A single counterexample can be used to show that A, B, and C are NOT always true. For example, setting $m = 4$ and $n = 1$ makes it clear that (m, n) need not necessarily satisfy any of A, B, and C, since $(4, 1) = 1$. Since the greatest common divisor of two different positive integers, m and n, must be less than or equal to the smaller of the two integers, it is clearly less than their product, mn, and the correct answer is D.

---

62. Which of the following can be used to develop the idea of similarity in geometry?

   I. Blueprints of floor plans
   II. Official road maps
   III. Scale models

   (A) I only
   (B) II only
   (C) III only
   (D) I and II only
   (E) I, II, and III

Similar figures must be identical in shape, corresponding angles must be congruent, and corresponding sides must be proportional. Because blueprints, official road maps, and scale models all have those characteristics, the best answer is E.

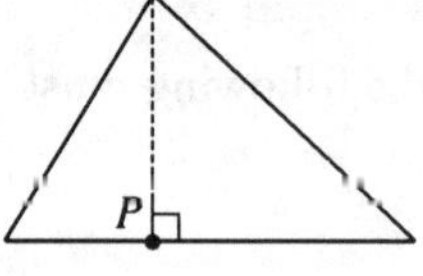 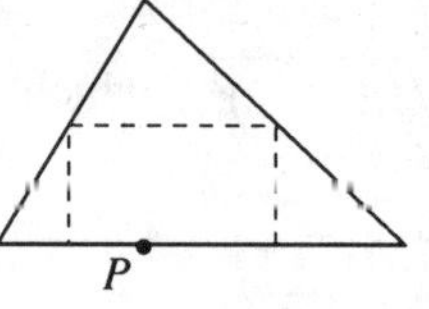 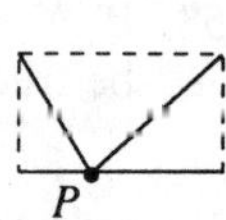

Figure I   Figure II   Figure III

63. In a geometry class, the teacher gives each student a triangular piece of paper marked as shown in Figure I. The students are then instructed to fold the triangle so that each vertex coincides with point P and a rectangle is formed by the dotted lines (folds), as shown in Figures II and III. Which of the following could the students discover from this demonstration?

   I. The sum of the measures of the angles of a triangle is equal to the measure of a straight angle.
   II. The length of the segment that joins the midpoints of two sides of a triangle is one-half the length of the third side.
   III. The area of the rectangular region formed as shown in Figure III is half the area of the original triangular region shown in Figure I.

   (A) I only
   (B) II only
   (C) III only
   (D) I and II only
   (E) I, II, and III

With respect to I, when the triangular piece of paper is folded along the dotted lines, as indicated in Figure II, the three vertices of the triangle come together at point P and together they form a straight angle whose rays are coincident with the side of the rectangle containing point P, as indicated in Figure III. Thus, the students should discover that the sum of the measures of the angles of a triangle is equal to the measure of a straight angle.

With respect to II, after the folds have been made, (1) the side of the triangle that contains point P (Figure I) has been placed, with double thickness, onto the side of the rectangle that contains point P (Figure III); and (2) the segment that joins the midpoints of the sides of the triangle that do not contain point P (Figure II) has been folded so it becomes the side of the rectangle (Figure III) opposite the side that contains point P. Since opposite sides of a rectangle are equal, the students should discover that the length of the segment that joins the midpoints of two sides of a triangle is one-half the length of the third side.

With respect to III, since the folds have transformed the triangle into a rectangle with double thickness, the students should discover that the area of the rectangular region formed in Figure III is half the area of the original triangular region shown in Figure I. Thus, the best answer is E.

64. For what values of x does the determinant
$$\begin{vmatrix} 1 & -x \\ -x & 1 \end{vmatrix} \text{ equal } 0 \text{ ?}$$

(A) No values of x
(B) $x = -1$ only
(C) $x = 0$
(D) $x = 1$ only
(E) $x = -1$ or $x = 1$

The value of the determinant $\begin{vmatrix} 1 & -x \\ -x & 1 \end{vmatrix}$ is equal to 0 if and only if

$$(1)(1) - (-x)(-x) = 0$$
$$1 - x^2 = 0$$
$$x^2 = 1$$
$$x = \pm 1.$$

Therefore, the correct answer is E.

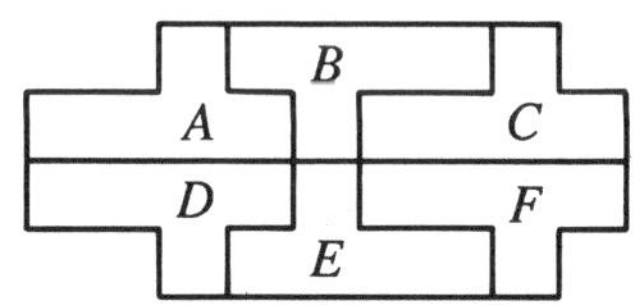

65. The figure above shows six identical tiles, all of which are red on one side and white on the other side. If tile A has the red side turned up, which of the other tiles have the red side up?

(A) B and C
(B) D and E
(C) B and D
(D) C and E
(E) C and F

Any tile that can slide onto A or be rotated and slid onto A is red. Therefore, tiles B and C have the red side up, and the correct answer is A.

---

66. If $\cos x \neq 0$, then $\dfrac{1 - \cos^2 x}{\cos^2 x} =$

(A) $\tan^2 x$

(B) $\sec^2 x$

(C) $\csc^2 x$

(D) $\cot^2 x$

(E) $\cos 2x$

$$\frac{1 - \cos^2 x}{\cos^2 x} = \frac{\sin^2 x}{\cos^2 x} = \left(\frac{\sin x}{\cos x} \right)^2 = \tan^2 x$$

Thus, the correct answer is A.

67. For the numbered statements below, the notation $p \leftarrow q$ is defined to mean "replace the value of p by the present value of q."

1. $x \leftarrow 5$, $y \leftarrow 3$
2. $z \leftarrow (x + y)$
3. Record the value of z.
4. $x \leftarrow (x + y)$
5. Go to step 2.

The <u>second</u> value of z recorded at step 3 will be

(A) 8
(B) 10
(C) 11
(D) 13
(E) 16

The results of following the successive instructions or steps are as follows:

Step 1: $x \leftarrow 5$, $y \leftarrow 3$

Step 2: $z \leftarrow 5 + 3$ $z \leftarrow 8 + 3$

Step 3: Record the value of z as 8.   Record the value of z as 11.

Step 4: $x \leftarrow 5 + 3$

Step 5: Go back to Step 2.

Since the second value of z recorded is 11, the correct answer is C.

---

68. If the line $5x - 12y = 60$ intersects the x-axis and the y-axis at points P and Q, what is the length of segment PQ ?

(A) 5
(B) 7
(C) 12
(D) 13
(E) 17

If $5x - 12y = 60$, then the x- and y-intercepts, P and Q, are $(12, 0)$ and $(0, -5)$, respectively. Then the length of PQ is

$$\sqrt{12^2 + (-5)^2} = \sqrt{169} = 13,$$ and the correct answer is D.

69. Which of the following describes the roots of
$-\dfrac{2x^2}{3} + \dfrac{5x}{7} + \dfrac{7}{10} = 0$?

(A) There is only one root.
(B) One is positive and one is negative.
(C) Both are positive.
(D) Both are negative.
(E) Both are imaginary.

If both sides of the equation given are multiplied by $-\dfrac{2}{3}$, the resulting equivalent equation will be $x^2 - \dfrac{15}{14}x - \dfrac{21}{20} = 0$. A second-degree equation has at most two real roots. Since the product of the roots, $-\dfrac{21}{20}$, is a negative real number, this second-degree equation has two roots, one of which is positive and one of which is negative. Therefore, the correct answer is B.

---

70. Which of the following graphs represents the solution set of $x^2 - 9 < 0$?

(A) 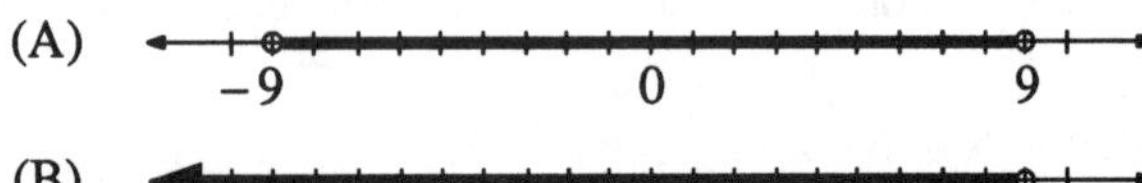

(B) 

If $x^2 - 9 < 0$, then $x^2 < 9$ and $|x| < \sqrt{9}$. Thus, $x < 3$ and $x > -3$. Since 3 and -3 are not in the solution set, the correct answer is D.

A cyclist traveled from Caldonia to Newtown at an average rate of 18 miles per hour and returned immediately by the same route at an average rate of 12 miles per hour. What was the average rate for the round trip?

71. With respect to the problem above, which of the following is true?

(A) The answer is 14.4 miles per hour.
(B) The answer is 15 miles per hour.
(C) The only additional information needed to determine the answer is the time for the round trip.
(D) The only additional information needed to determine the answer is the distance between the towns.
(E) Both the total time and the distance are needed to determine the answer.

If Caldonia and Newtown are D miles apart, it took $\dfrac{D}{18}$ hours to go from Caldonia to Newtown and $\dfrac{D}{12}$ hours for the return trip. The average rate for the round trip is $\dfrac{\text{total distance traveled in miles}}{\text{hours spent traveling}}$

$$= \dfrac{2D}{\dfrac{D}{18} + \dfrac{D}{12}} = \dfrac{2}{\dfrac{1}{18} + \dfrac{1}{12}} = \dfrac{2}{\dfrac{2}{36} + \dfrac{3}{36}} = \dfrac{2}{\dfrac{5}{36}} = \dfrac{72}{5} = 14.4$$

miles per hour. Thus, the average rate for the round trip was 14.4 miles per hour, and the correct answer is A.

72. A plane intersects a right circular cylinder (with two bases) and is parallel to the axis of the cylinder. If the plane passes through the interior of the cylinder, then the intersection of the plane with the surface of the cylinder is

(A) exactly 4 points
(B) a pair of parallel line segments
(C) a rectangle
(D) a circle
(E) an ellipse

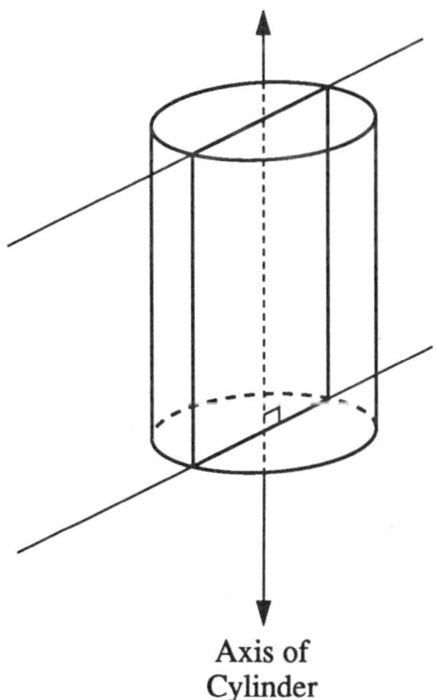

The figure above shows that the desired intersection is a rectangle. More formally, since the plane is parallel to the axis of the right circular cylinder, it must be perpendicular to both bases of the cylinder and must intersect the bases in parallel, congruent line segments. Furthermore, since the plane is parallel to the axis, it is also parallel to every altitude of the cylinder. Since opposite sides of the intersection are parallel and congruent and the base intersections are perpendicular to the lateral intersections, the figure must be a rectangle. The correct answer is C.

73. If f is a function such that $f(3) = 2$, $f(4) = 2$, and $f(n + 4) = f(n + 3)f(n + 2)$ for all integers $n \geqq 0$, what is the value of $f(6)$?

(A) 4
(B) 5
(C) 6
(D) 8
(E) It cannot be determined from the information given.

If $f(n+4) = f(n+3)f(n+2)$ for all integers $n \geq 0$, then $f(n+2) = f(n+1)f(n)$ for all integers $n \geq 2$. Thus, $f(5) = f(4) \cdot f(3) = 2 \cdot 2 = 4$ and $f(6) = f(5) \cdot f(4) = 4 \cdot 2 = 8$. Therefore, the correct answer is D.

___

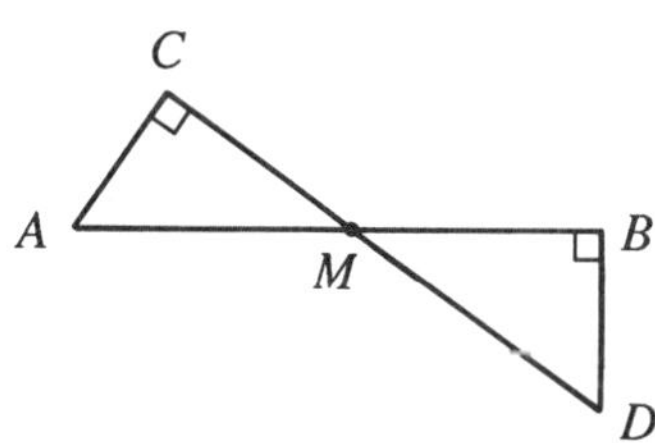

74. In the figure above, if $\overline{AB}$ and $\overline{CD}$ intersect at M, and if $\angle B$ and $\angle C$ are right angles, then $\dfrac{MB}{BD} =$

(A) $\dfrac{MC}{CA}$

(B) $\dfrac{MA}{AC}$

(C) $\dfrac{MC}{MA}$

(D) $\dfrac{MD}{MA}$

(E) $\dfrac{MA}{MC}$

Since all right angles are congruent and vertical angles are congruent, it follows that $\angle C$ is congruent to $\angle B$ and $\angle AMC$ is congruent to $\angle BMD$. If two angles of a triangle are congruent to two angles of a second triangle, the two triangles are similar. Thus $\triangle ACM$ is similar to $\triangle DBM$ and $\angle MDB$ is congruent to $\angle MAC$. If two triangles are similar, then their corresponding sides are proportional. (The corresponding sides are the sides opposite the congruent angles.) Therefore, $\dfrac{MA}{MD} = \dfrac{CA}{BD} = \dfrac{MC}{MB}$ and, consequently, $\dfrac{MB}{BD} = \dfrac{MC}{CA}$. The correct answer is A.

75. If each of the acute angles of a parallelogram has degree measure 60, and the lengths of the sides are 3 and 4, then the area of the parallelogram is

(A) 6

(B) $4\sqrt{3}$

(C) $4\sqrt{6}$

(D) $6\sqrt{3}$

(E) 12

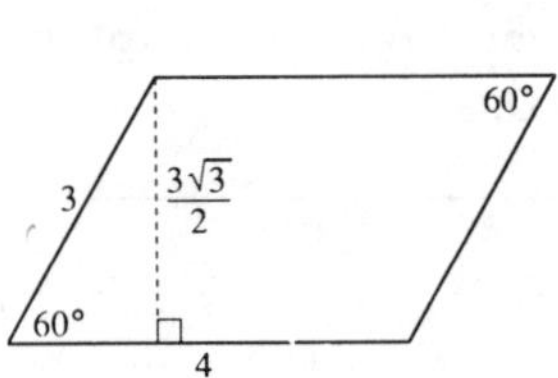

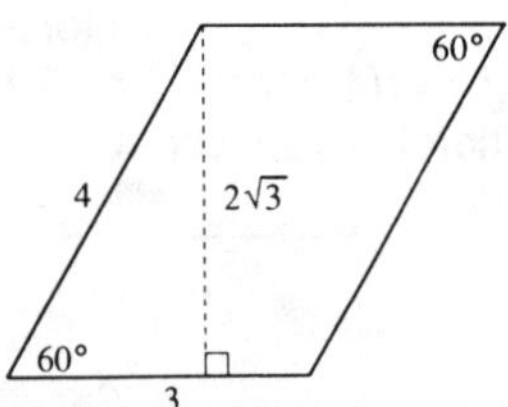

The parallelogram must have dimensions as indicated in one or the other of the figures shown above. In the first figure, the length of the base of the parallelogram is 4 and its altitude is $\frac{3\sqrt{3}}{2}$. (Note that the triangles formed by the altitudes and the sides of the respective parallelograms have degree measures 30, 60, and 90.) In the second figure, the length of the base of the parallelogram is 3 and its altitude is $2\sqrt{3}$. In either case, the area of the parallelogram is

$4\left(\dfrac{3\sqrt{3}}{2}\right) = 2(3\sqrt{3}) = 6\sqrt{3}$ or $3(2\sqrt{3}) = 6\sqrt{3}$. The correct answer is D.

---

76. If $f(x) = \dfrac{2x-5}{2}$ and $g(x) = \dfrac{2}{2x-5}$, then $g(f(10)) =$

(A) $\dfrac{1}{10}$

(B) $\dfrac{1}{5}$

(C) 1

(D) 5

(E) 10

To find $g(f(10))$, first find $f(10)$:

$$f(10) = \frac{2(10)-5}{2} = 7.5$$

Now, $g(f(10)) = g(7.5) = \dfrac{2}{2(7.5)-5} = \dfrac{2}{10} = \dfrac{1}{5}$. The correct answer is B.

77. S is the set of all numbers which can be written in the form $2^n 3^m$ where n and m are positive integers. If a and b are two numbers in S, which of the following must also be in S?

(A) $a + b$

(B) $a - b$

(C) $\sqrt{ab}$

(D) $\dfrac{a}{b}$

(E) ab

To show that any one of the numbers listed as an option is not necessarily in set S, it is sufficient to find a single counterexample. One counterexample that rules out choices A, B, C, and D is $a = 48$, which has the factors $2^4 3^1$, and $b = 18$, which has the factors $2^1 3^2$. Although both a and b are in S, none of the numbers $48+18$, $48-18$, $\sqrt{48(18)}$, and $\dfrac{48}{18}$ is of the form $2^n 3^m$, and consequently, none is in set S. With respect to choice E, if $a = 2^m 3^n$ and $b = 2^x 3^y$, where m, n, x, and y are all positive integers, then $ab = 2^m 3^n (2^x 3^y) = (2^{m+x})(3^{n+y})$, which must be in set S. The correct answer is E.

---

78. Each of the following defines a function such that the interval $[0, 1]$ is mapped into the interval $[0, 1]$ EXCEPT

(A) $f(x) = \sqrt{x}$

(B) $f(x) = |x|$

(C) $f(x) = \dfrac{1}{x}$

(D) $f(x) = x^2$

(E) $f(x) = x^3$

Showing that a function f does not map $[0, 1]$ into $[0, 1]$ is the same as showing that for some x in $[0, 1]$, $f(x)$ is not in $[0, 1]$ (that is, for some x in $[0, 1]$, either $f(x) < 0$ or $f(x) > 1$). For all x in $[0, 1]$, $\sqrt{x}, |x|, x^2$, and x^3 are in $[0, 1]$. Thus, none of the choices A, B, D, and E is the correct answer. With respect to choice C, $f(0)$ is not defined and $f(x) > 1$ for every x in $[0, 1]$. Therefore, $f(x) = \dfrac{1}{x}$ does NOT map $[0, 1]$ into $[0, 1]$, and C is the correct answer.

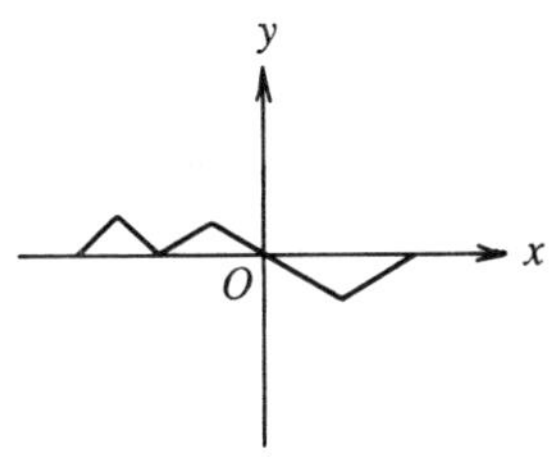

79. The figure above shows the graph of $y = f(x)$. Which of the following is the graph of $y = -f(x)$?

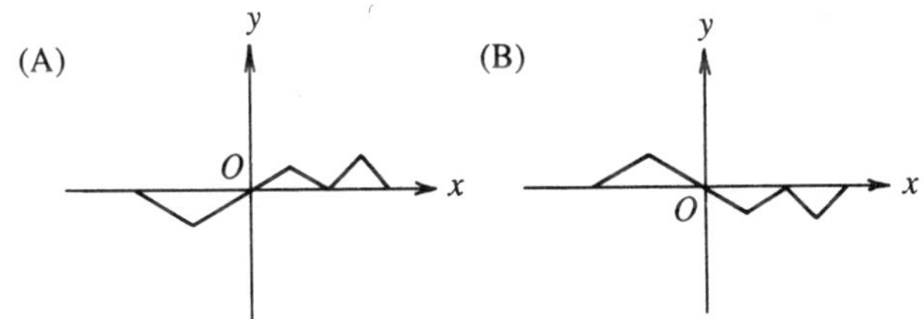

(A)   (B)

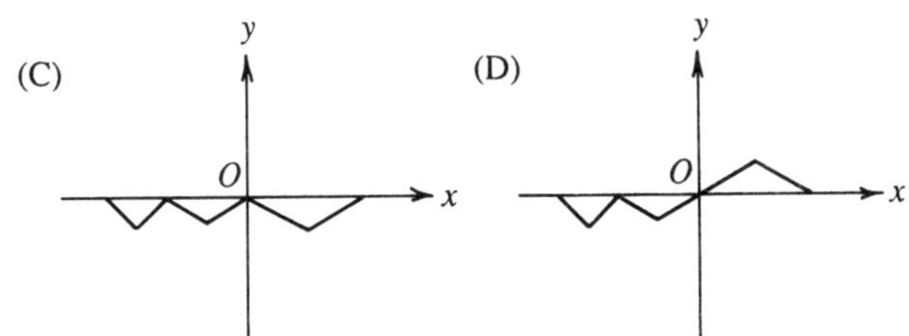

(C)   (D)

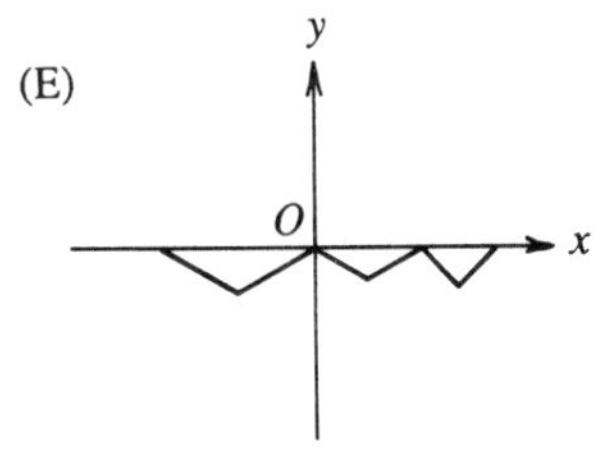

(E)

A point on the graph of the function $y = f(x)$ has coordinates $(x, f(x))$, and a point on the graph of the function $y = -f(x)$ has coordinates $(x, -f(x))$. Now both functions are defined for the same values of x, and for every value of x, $-f(x)$ is the negative of $f(x)$; therefore, the graph of $y = -f(x)$ is a reflection in the x-axis of the graph of $y = f(x)$. Only choice D shows the graph of a reflection of the function $y = f(x)$ in the x-axis. The correct answer is D.

80. If the results of tossing a fair coin 5 times are 5 heads in a row, what is the probability of getting a head on the 6th toss?

(A) 0

(B) $\left(\dfrac{1}{2}\right)^6$

(C) $\dfrac{1}{6}$

(D) $\dfrac{1}{2}$

(E) 1

Each toss of a coin is independent of all other tosses; thus, the results of the 5 tosses do not affect the probability of getting a head on the 6th toss. Since the coin is given to be a fair coin, the probability of getting a head on any toss, including the 6th toss, is $\dfrac{1}{2}$. The correct answer is D.

81. What is the area of the region bounded below by the graph of $y = x^2 - 4$ and bounded above by the graph of $y = 4 - x^2$?

(A) 0

(B) $\dfrac{16}{3}$

(C) $\dfrac{32}{3}$

(D) $\dfrac{64}{3}$

(E) $\dfrac{128}{3}$

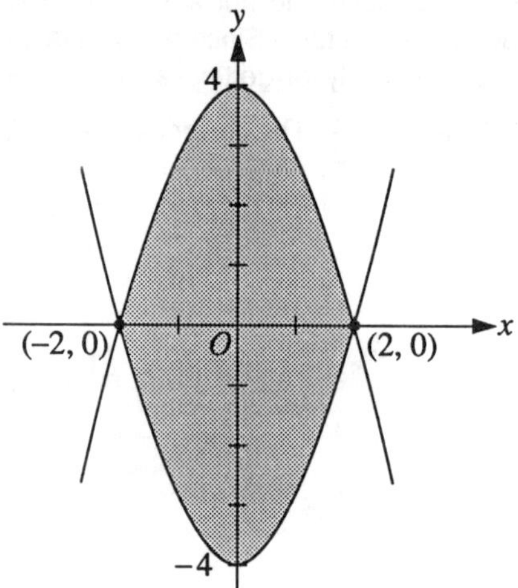

The shaded region in the figure above is the region bounded below by the graph of $y = x^2 - 4$ and above by the graph $y = 4 - x^2$. Note that the two graphs intersect at the values of x where $x^2 - 4 = 4 - x^2$ and that the enclosed region is symmetric with respect to both the x-axis and the y-axis. Because of this symmetry, the area of the entire region can be found by finding the area of the portion in quadrant I and then multiplying that area by 4. Thus, the area is given by the integral

$$4\int_0^2 (4 - x^2)\,dx = 4\left[(4x - \frac{x^3}{3})\right]_0^2$$
$$= 4\left[(8 - \frac{8}{3}) - (0 - 0)\right]$$
$$= 4\left[8 - \frac{8}{3}\right] = \frac{64}{3}.$$

The correct answer is D.

82. If length L is measured to be 7.24 meters and if this measurement is accurate to the nearest hundredth of a meter, then the greatest possible error in this measurement is

(A) 0.0005 m
(B) 0.005 m
(C) 0.05 m
(D) 0.5 m
(E) 1.0 m

If the measurement 7.24 meters is accurate to the nearest hundredth of a meter, then $7.235 \le L < 7.245$. The greatest possible error in this measurement would occur if the true length were 7.235 and it were rounded up to 7.24. The amount of error would then be $7.24 - 7.235 = 0.005$ meter. The correct answer is B.

---

83. In trying to prove that $\sqrt{3}$ is irrational using the indirect method of proof, one should do which of the following?

(A) Assume that $\sqrt{3}$ is rational and show that this leads to the contradiction of a known fact.

(B) Assume that $\sqrt{3}$ is rational and show that this does not lead to the contradiction of a known fact.

(C) Assume that $\sqrt{3}$ is not rational and show that this leads to the contradiction of a known fact.

(D) Assume that $\sqrt{3}$ is not rational and show that this does not lead to the contradiction of a known fact.

(E) None of the above, since one cannot prove by the indirect method that $\sqrt{3}$ is irrational.

A theorem can be proved indirectly by assuming that its conclusion is false and then showing that this assumption leads to the contradiction of a known fact. To prove that $\sqrt{3}$ is irrational, assume $\sqrt{3}$ is rational. Then $\sqrt{3} = \dfrac{a}{b}$, where a and b are positive integers that have no common factor greater than 1.

Now $\sqrt{3} = \dfrac{a}{b}$ implies that $3 = \dfrac{a^2}{b^2}$, which implies that $3b^2 = a^2$. Thus, 3 divides a^2 implies that 3 divides a. So $a = 3c$, where c is a positive integer. Then $3b^2 = a^2 = (3c)^2 = 9c^2$ implies that $b^2 = 3c^2$. By the same reasoning as before, 3 divides b. It has been shown that 3 divides both a and b, but the fraction $\dfrac{a}{b}$ was assumed to be in lowest terms, which is a contradiction. Thus, $\sqrt{3}$ is irrational. The correct answer is A. Choice B is incorrect because, if assuming the conclusion is false does not lead to a contradiction, the indirect method of proof does not work. Choices C and D are incorrect because, in each case, one is assuming the conclusion is true. Since the desired proof is shown above, obviously E is not correct.

84. The complex numbers satisfy all of the following
EXCEPT

(A) the order properties
(B) closure under raising to a power
(C) associativity under multiplication
(D) the group properties for addition
(E) the distributive property

Recall that the complex numbers are an algebraically closed field with respect to the standard operations of addition and multiplication. Because choices B, C, D, and E are properties of an algebraically closed field, the complex numbers satisfy these properties. The only choice remaining is A, which is the correct answer. To prove that the complex numbers do not satisfy the order properties, recall that set S can be ordered if there exists a relation $(>)$ on S that satisfies the following axioms for every x, y, and z in S.

1. Exactly one of the relations $x = y$, $x \;(>)\; y$, or $y \;(>)\; x$ holds.

2. If $x \;(>)\; y$, then $x + z \;(>)\; y + z$.

3. If $x \;(>)\; 0$ and $y \;(>)\; 0$, then $xy \;(>)\; 0$.

4. If $x \;(>)\; y$ and $y \;(>)\; z$, then $x \;(>)\; z$.

Assume there is a relation $(>)$ on the complex numbers that satisfies these axioms. Then either $i \;(>)\; 0$ or $0 \;(>)\; i$ by axiom 1.

Case I: $i \;(>)\; 0$

$i \;(>)\; 0$ implies that $i \cdot i = -1 \;(>)\; 0$ (axiom 3), which implies that $(-1)(i) = -i \;(>)\; 0$ (axiom 3), which implies that $-i + i \;(>)\; 0 + i$ (axiom 2), which implies that $0 \;(>)\; i$. Thus, $i \;(>)\; 0$ and $0 \;(>)\; i$, which is impossible by axiom 1. Therefore, $i \;(\not>)\; 0$.

Case II: $0 \;(>)\; i$

$0 \;(>)\; i$ implies that $0 - i \;(>)\; i - i$ (axiom 2), which implies that $-i \;(>)\; 0$ and $(-i)(-i) = -1 \;(>)\; 0$ (axiom 3), which implies that $(-1)(-i) = i \;(>)\; 0$ (axiom 3). Thus, $0 \;(>)\; i$ and $i \;(>)\; 0$, which is impossible by axiom 1. Therefore, $0 \;(\not>)\; i$. Since $0 \;(\not>)\; i$, $i \;(\not>)\; 0$, and clearly $i \neq 0$, it follows that the complex numbers do not satisfy the order properties.

85. Let G be a group with operation $*$. If a, b, and c are elements in G, then which of the following statements is NOT necessarily true?

(A) $a * b$ is in G
(B) $a * (b * c) = (a * b) * c$
(C) $a * b = b * a$
(D) There exists an element e in G such that $a * e = e * a = a$ for each a in G.
(E) For each a in G there exists an element a^{-1} in G such that $a * a^{-1} = e$.

A nonempty set of elements G together with operation $*$ is said to form a group if those properties exemplified in choices A, B, D, and E are satisfied. These properties include closure, associativity, identity element, and an inverse for every element in G. With respect to choice C, a group need not be commutative. Since C is not necessarily true, the correct answer is C.

---

86. If f is a trigonometric function whose graph has positive slope wherever it is defined, then $f(x) =$

(A) $\sin x$
(B) $\cos x$
(C) $\tan x$
(D) $\sec x$
(E) $\cot x$

The slopes of the graphs of $\sin x$, $\cos x$, and $\sec x$ change back and forth from positive to negative, with the slope being 0 at the local maxima and minima. The graph of $\tan x$ has positive slope wherever the function is defined, and the graph of $\cot x$ has negative slope wherever the function is defined. The correct answer is C.

87. What are all possible values of x for which
$\cos^2 x = \cos x$ and $0 \leq x < 2\pi$?

(A) 0 only

(B) $\dfrac{3\pi}{2}$ only

(C) 0 and $\dfrac{\pi}{2}$ only

(D) $\dfrac{\pi}{2}$ and $\dfrac{3\pi}{2}$ only

(E) 0, $\dfrac{\pi}{2}$, and $\dfrac{3\pi}{2}$

If $\cos^2 x = \cos x$, then $\cos^2 x - \cos x = 0$,
$\cos x(\cos x - 1) = 0$, which implies that
$\cos x = 0$ or $\cos x = 1$.

One way to locate the values of x for which $\cos x = 0$ or 1 in the interval $0 \leq x < 2\pi$ is to look at the graph of $\cos x$. In the figure below, note that $\cos x = 0$ when $x = \dfrac{\pi}{2}$ and $x = \dfrac{3\pi}{2}$; $\cos x = 1$ when $x = 0$ and 2π. Since 2π is not in the desired interval, the correct values of x are 0, $\dfrac{\pi}{2}$, and $\dfrac{3\pi}{2}$. The correct answer is E.

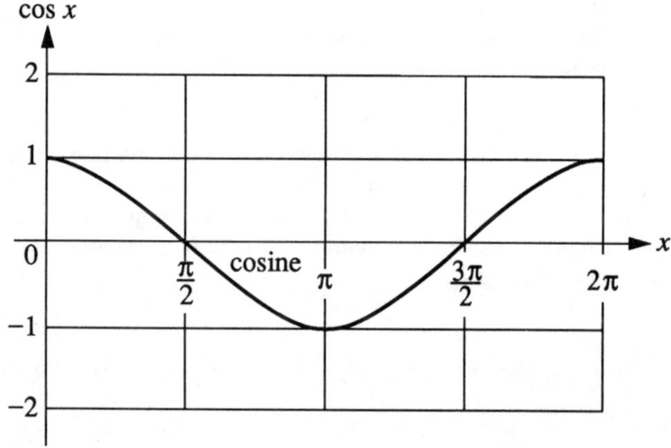

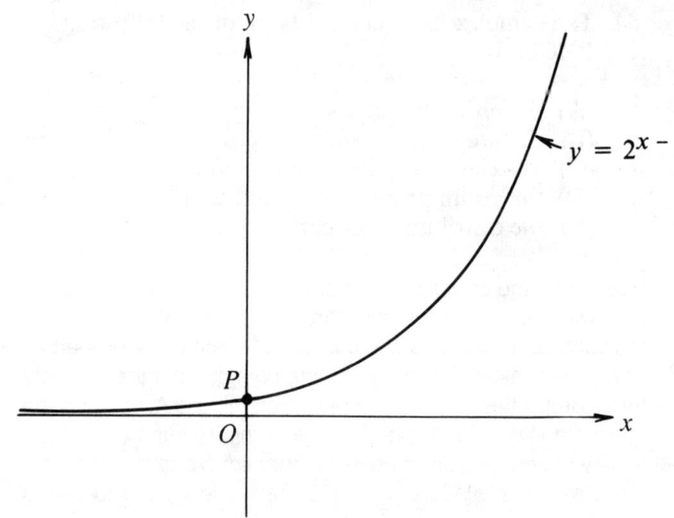

88. The figure above shows the graph of $y = 2^{x-2}$.
The point P is

(A) $\left(\dfrac{1}{4}, 0\right)$

(B) $(1, 0)$

(C) $(0, 1)$

(D) $\left(0, \dfrac{1}{2}\right)$

(E) $\left(0, \dfrac{1}{4}\right)$

Since the point P is on the y-axis, it has coordinates $(0,k)$. Therefore, the graph of the exponential function $y = 2^{x-2}$ passes through point P when $x = 0$. When $x = 0$, $y = 2^{-2} = \dfrac{1}{4}$. Thus, the coordinates of P are $\left(0, \dfrac{1}{4}\right)$. The correct answer is E.

89. Which of the following do NOT necessarily determine a plane?

(A) A line and a point not on the line
(B) Two distinct parallel lines
(C) The vertices of a triangle
(D) Two distinct intersecting lines
(E) Three distinct points

At least three noncollinear points are needed to determine a plane. Note that choices A, B, C, and D give conditions that include at least three noncollinear points, whereas choice E does not specify that the points are noncollinear. The correct answer is E.

90. If y varies directly as x^2, and if $y = 1$ when $x = 0.1$, for what values of x does $y = \frac{1}{2}$?

(A) $\pm\dfrac{1}{200}$

(B) $\pm\dfrac{1}{20}$

(C) $\pm\dfrac{\sqrt{2}}{20}$

(D) $\pm\sqrt{5}$

(E) $\pm 5\sqrt{2}$

If y varies directly as x^2, then $y = kx^2$ for some constant k. If $y = 1$ when $x = 0.1$, then $1 = k(0.1)^2$ and $k = 100$. To find the values of x for which $y = \dfrac{1}{2}$, it is necessary to substitute $y = \dfrac{1}{2}$ into the equation $y = 100x^2$. Thus, $\dfrac{1}{2} = 100x^2$ or $x^2 = \dfrac{1}{200}$. Therefore, $x = \pm\dfrac{\sqrt{2}}{20}$. The correct answer is C.

---

91. Let f be the function given by $f(x) = 2x - 1$. The inverse function of f is given by $f^{-1}(x) =$

(A) $1 - 2x$

(B) $\dfrac{1}{2x - 1}$

(C) $\dfrac{x + 1}{2}$

(D) $2x + 1$

(E) $\dfrac{x - 1}{2}$

If f is a one-to-one function, then the inverse function of f, denoted by f^{-1}, is that function for which $f(f^{-1}(x)) = x$ for all x in the domain of f^{-1} and $f^{-1}(f(x)) = x$ for all x in the domain of f. The inverse function, $f^{-1}(x)$, for $f(x) = 2x - 1$ can be found by solving the equation $f(x) = 2x - 1$ for x in terms of $f(x)$ and then replacing $f(x)$ by x. Therefore, $f(x) = 2x - 1$ implies that $2x = f(x) + 1$ and $x = \dfrac{f(x) + 1}{2}$. Thus, $f^{-1}(x) = \dfrac{x + 1}{2}$. The correct answer is C.

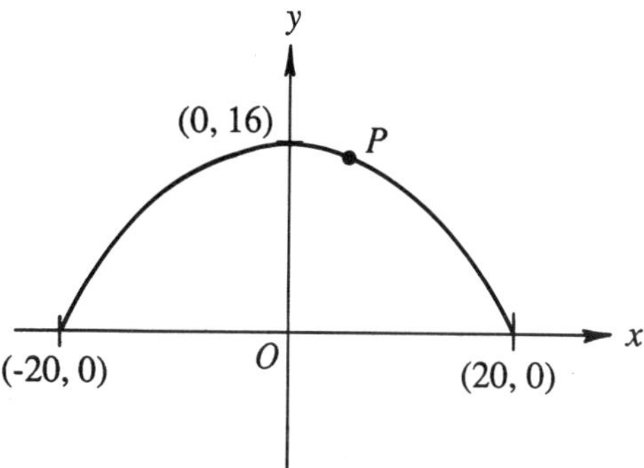

92. The figure above indicates the coordinates of three points on the parabolic arc. If P is the point on the arc that has coordinates $(5, \ y)$, then $y =$

(A) 12

(B) 15

(C) $15\dfrac{1}{3}$

(D) $15\dfrac{1}{2}$

(E) $15\dfrac{3}{4}$

Since the parabola in the figure is symmetric with respect to the y-axis, its equation is of the form $y = ax^2 + c$. Note from the graph that when $x = 0$, $y = c = 16$ and when $y = 0$, $x = \pm 20$. Substituting these values into the equation $y = ax^2 + 16$ gives $0 = a(20)^2 + 16$, $400a = -16$, and $a = -\dfrac{1}{25}$. Thus, the equation for this parabola is $y = -\dfrac{1}{25}x^2 + 16$. To find the y coordinate of point P, it is necessary to evaluate the equation when $x = 5$. Thus, $y = -\dfrac{1}{25}(5)^2 + 16 = 15$. The correct answer is B.

$$\begin{cases} 3^{x+y} = 81 \\ 36^{\frac{y}{2}} = 216 \end{cases}$$

93. In the system of equations above, $x =$

(A) 0
(B) 1
(C) 2
(D) 3
(E) 4

From the equation $3^{x+y} = 81$, it can be determined that $x + y = 4$ since $3^4 = 81$. From the equation $36^{\frac{y}{2}} = 216$, it can be determined that $(6^2)^{\frac{y}{2}} = 6^y = 216$, and $y = 3$. If $x + y = 4$ and $y = 3$, then $x = 1$. The correct answer is B.

94. The slope of a curve at every point (x, y) on the curve is given by $10x - 1$. If the curve contains the point $(1, 5)$, then an equation of the curve is

(A) $y = 10x^2 + x - 1$
(B) $y = 5x^2 + x + 1$
(C) $y = 5x^2 - x$
(D) $y = 5x^2 - x + 1$
(E) $y = 5x^2 + x$

If $y = f(x)$ is the equation for a certain curve, and f is a differentiable function, then the first derivative of $f(x)$ gives the slope of the curve at every point (x,y) on the curve. Since it is given that the slope, $f'(x) = 10x - 1$, the equation of the curve is $f(x) = \int (10x - 1)\, dx = 5x^2 - x + c$. If the curve contains the point $(1,5)$, then $5 = 5(1)^2 - 1 + c$, and $c = 1$. Thus, the equation of the curve is $y = 5x^2 - x + 1$. The correct answer is D.

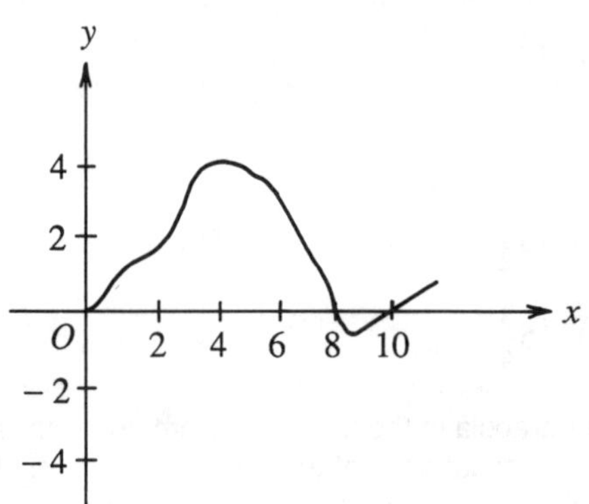

95. For all x in the interval $[0, 10]$, F is the function defined by $F(x) = \int_0^x f$, where f is the function that has the graph shown above. At what value of x does F attain its maximum value?

(A) 0
(B) 4
(C) 6
(D) 8
(E) 10

If a function $f(x)$ is nonnegative on the interval $[a,b]$, then $\int_a^b f(x)\, dx$ represents the area of the region bounded by $y = f(x)$, the x-axis, and the lines $x = a$ and $x = b$. On the other hand, if $f(x)$ is negative on the interval $[a,b]$, then $\int_a^b f(x)\, dx$ represents the negative of the area of the region bounded by $y = f(x)$, the x-axis, and the lines $x = a$ and $x = b$. Therefore, the function $F(x)$ increases when $f(x) > 0$ and decreases when $f(x) < 0$. Since $f(x) > 0$ when $x < 8$ and $f(x) < 0$ when $x > 8$, it follows that $F(x)$ attains its maximum value when $x = 8$. The correct answer is D.

96. Which of the following could NOT be used as a definition of the Euclidean distance between any two distinct points x_1 and x_2 on the number line?

(A) $\begin{cases} x_2 - x_1, & \text{if } x_2 > x_1 \\ x_1 - x_2, & \text{if } x_1 > x_2 \end{cases}$

(B) $\left| \int_{x_1}^{x_2} dx \right|$

(C) $\sqrt{(x_1 - x_2)^2}$

(D) $x_1 - x_2$

(E) $\max\{(x_1 - x_2), (x_2 - x_1)\}$

Note that choices A, B, C, and E are all ways of expressing $|x_1 - x_2|$, whereas choice D is negative if $x_2 > x_1$. Since the distance between two points can never be negative, D is the correct answer.

---

97. A teacher instructed a student to factor completely the polynomial $x^3 + 2x^2 - 5x - 10$. The teacher expected the answer $(x^2 - 5)(x + 2)$, but the student wrote $(x + \sqrt{5})(x - \sqrt{5})(x + 2)$. Which of the following instructions would have indicated more precisely what the teacher expected?

   I. Factor into a product of primes.
   II. Factor completely over the integers.
   III. Factor into polynomials that have integer coefficients and are of the lowest possible degree.

(A) I only
(B) II only
(C) III only
(D) I or III
(E) II or III

With respect to instruction I, a prime polynomial is a polynomial with coefficients in a ring R that cannot be expressed as a product of polynomials, each of lower degree, with coefficients in that ring. Note that this definition depends on the ring R, which is not specified in instruction I. If R is the ring of integers, the prime factorization would be $(x^2 - 5)(x + 2)$, and if R is the ring of real numbers, the prime factorization would be $(x + \sqrt{5})(x - \sqrt{5})(x + 2)$. Thus, instruction I would not have indicated more precisely what the teacher expected. Instructions II and III are equivalent and indicate that the coefficients of the factors cannot be irrational, thus ruling out the factorization $(x + \sqrt{5})(x - \sqrt{5})(x - 2)$, but allowing the factorization $(x^2 - 5)(x + 2)$. This is precisely what the teacher expected. Thus, the best answer is E.

98. An algorithm for multiplying decimals is as follows:

1. Treat the numbers as whole numbers and multiply.
2. In the result, point off from right to left the total number of decimal places in the two factors.

This algorithm is based on which of the following?

(A) $\dfrac{a^n}{10^n} \times \dfrac{b^n}{10^n} = \dfrac{(ab)^n}{10^{2n}}$

(B) $\dfrac{a}{10^m} \times \dfrac{b}{10^n} = \dfrac{ab}{10^{m+n}}$

(C) $a^n(b^n + c^n) = (ab)^n + (ac)^n$

(D) If $ab = c$, then $\dfrac{c}{a} = b$.

(E) If $\dfrac{c}{a} = b$, then $ab = c$.

This algorithm assumes that each of the decimals being multiplied is a terminating decimal since the instructions could not be implemented if the decimals were nonterminating decimals. Any two such decimals can be written in fraction form as $\dfrac{a}{10^m}$ and $\dfrac{b}{10^n}$, where a and b are integers. For example, 5.235 can be written as $\dfrac{5235}{10^3}$ and 75.62 can be written as $\dfrac{7562}{10^2}$. Then, using the algorithm for multiplying fractions, $(5.235)(75.62) = \dfrac{5235}{10^3} \times \dfrac{7562}{10^2} = \dfrac{5235 \times 7562}{10^{3+2}}$ or, more generally, $\dfrac{a}{10^m} \times \dfrac{b}{10^n} = \dfrac{a \times b}{10^{m+n}}$. This is equivalent to multiplying the two integers, a and b, and then pointing off from right to left $m + n$ decimal places. Note that the general case is precisely what is given as choice B. Although the remaining options represent true statements, only B is a basis for the algorithm for multiplying two decimals.

99. If a box of 2 dozen ballpoint pens contains 5 that are defective, what is the probability that 2 pens both selected at random and without replacement will be defective?

(A) $\left(\dfrac{5}{24}\right)\left(\dfrac{4}{23}\right)$

(B) $\left(\dfrac{5}{24}\right)\left(\dfrac{4}{24}\right)$

(C) $\left(\dfrac{19}{24}\right)\left(\dfrac{19}{23}\right)$

(D) $\left(\dfrac{5}{24}\right)^2$

(E) $\left(\dfrac{19}{24}\right)^2$

The probability of a desired outcome to an experiment is equal to the ratio of the total number of possible desirable outcomes to the total number of possible outcomes. Thus, the probability of selecting a defective pen in the first selection is $\dfrac{5}{24}$. If the first pen selected is defective, there are 23 pens left, 4 of which are defective. The probability of selecting a defective pen under these conditions is $\dfrac{4}{23}$. The probability of both pens selected being defective is the product of these two probabilities, or $\left(\dfrac{5}{24}\right)\left(\dfrac{4}{23}\right)$. Therefore, the correct answer is A.

---

100. If I is the set of all integers and $x \in I$, which of the following denotes a one-to-one mapping of I <u>onto</u> I?

(A) $x \to x - 4$
(B) $x \to 2x + 1$
(C) $x \to 2x$
(D) $x \to x^2$
(E) $x \to x^3$

A mapping $f: I \to I$ is one-to-one and onto if, for every $y \in I$, there is a unique x satisfying $f(x) = y$. With respect to options B, C, D, and E, the mappings are not onto since the mappings yield only odd numbers, even numbers, perfect squares, and perfect cubes, respectively. With respect to choice A, the mapping takes any integer and decreases it by 4; clearly it is both one-to-one and onto. Thus, the correct answer is A.

101. The statement, "You will be healthy only if you exercise regularly," is equivalent to which of the following?

    I. If you exercise regularly, then you will be healthy.
    II. If you are healthy, then you exercise regularly.
    III. A sufficient condition for being healthy is that you exercise regularly.

(A) I only
(B) II only
(C) I and II only
(D) I and III only
(E) I, II, and III

The statement, "You will be healthy only if you exercise regularly," is equivalent to Statement II, "If you are healthy, then you exercise regularly." If either statement is true, both are true. Statements I and III are equivalent, and each is the converse of Statement II; these statements may or may not be true. Since only Statement II is equivalent to the given statement, the correct answer is B.

---

102. What is $\lim\limits_{x \to 4} \dfrac{x^3 - 64}{x^2 - 16}$?

(A) 0
(B) 1
(C) 6
(D) 12
(E) The limit does not exist.

Since direct substitution of $x = 4$ in $\dfrac{x^3 - 64}{x^2 - 16}$ in the calculation of $\lim\limits_{x \to 4} \dfrac{x^3 - 64}{x^2 - 16}$ yields the indeterminate form $\dfrac{0}{0}$, one can apply L'Hopital's rule, $\left(\lim\limits_{x \to a} \dfrac{f(x)}{g(x)} = \lim\limits_{x \to a} \dfrac{f'(x)}{g'(x)} \right)$, to calculate the limit. Therefore, $\lim\limits_{x \to 4} \dfrac{x^3 - 64}{x^2 - 16} = \lim\limits_{x \to 4} \dfrac{3x^2}{2x} = \dfrac{3(4)^2}{2(4)} = 6$. The correct answer is C.

---

103. For which of the following topics would a hand calculator be LEAST useful as a teaching device?

(A) Finding the limit of a function
(B) Finding the length of an unknown side of a triangle by the use of trigonometry
(C) Finding the coordinates of points that satisfy an equation
(D) Finding the mean of a frequency distribution
(E) Proving a theorem by mathematical induction

Each of the first four tasks listed as options could require considerable computation, whereas proving a theorem by mathematical induction would require only logic and algebraic manipulation for which a hand calculator would be of little, if any, use. The best answer is E.

104. If right circular cylinder A has a radius twice that of right circular cylinder B and a height 3 times that of B, then the volume of A is how many times the volume of B?

(A) 2
(B) 3
(C) 5
(D) 6
(E) 12

If the radius and height of cylinder B are r and h, respectively, then the volume is $\pi r^2 h$. The volume of cylinder A would then be $\pi(2r)^2(3h) = \pi(4r^2)(3h) = 12(\pi r^2 h)$. Thus, the volume of cylinder A is 12 times the volume of cylinder B. The correct answer is E.

---

105. Which of the following series converges?

(A) $\dfrac{1}{2} + \dfrac{2}{3} + \dfrac{3}{4} + \dfrac{4}{5} + \dfrac{5}{6} + \ldots + \dfrac{n}{n+1} + \ldots$

(B) $1 + \dfrac{2}{\sqrt{2}} + \dfrac{3}{\sqrt{3}} + \dfrac{4}{\sqrt{4}} + \dfrac{5}{\sqrt{5}} + \ldots + \dfrac{n}{\sqrt{n}} + \ldots$

(C) $\dfrac{2}{2} + \dfrac{2}{3} + \dfrac{2}{4} + \dfrac{2}{5} + \ldots + \dfrac{2}{n+1} + \ldots$

(D) $1 + \dfrac{1}{2} + \dfrac{1}{3} + \dfrac{1}{4} + \ldots + \dfrac{1}{n} + \ldots$

(E) $1 + \dfrac{1}{3} + \dfrac{1}{9} + \dfrac{1}{27} + \dfrac{1}{81} + \ldots + \dfrac{1}{3^{n-1}} + \ldots$

A necessary, though not sufficient, condition for a series $\{a_n\}$ to converge is $\lim\limits_{n \to \infty} a_n = 0$. With respect to choice A, $\lim\limits_{n \to \infty} \dfrac{n}{n+1} = \lim\limits_{n \to \infty} \dfrac{1}{1 + \frac{1}{n}} = 1$. With respect to choice B, $\lim\limits_{n \to \infty} \dfrac{n}{\sqrt{n}} = \lim\limits_{n \to \infty} \sqrt{n} = \infty$. Therefore, neither of these two series converges. Since $\lim\limits_{n \to \infty} a_n = 0$ for each of the remaining choices, it is necessary to examine these more carefully. If choice D, the harmonic series, is S, then choice C is $1 + 2S$, so that either both of these converge or both diverge. Since the harmonic series is most often used as an example of a divergent series in textbooks, the proof that it is divergent will not be developed here. Choice E is a geometric series with $r = \dfrac{1}{3}$. Since a geometric series converges whenever $|r| < 1$ and diverges otherwise, it follows that this is a convergent series. Thus, E is the only one of the five series given that converges.

106. If the sequence a_n is defined by

$$a_n = \left[2 + \frac{(-1)^n}{n}\right], \text{ then } \lim_{n \to \infty} a_n \text{ is}$$

(A) $-\infty$
(B) -2
(C) 0
(D) 2
(E) ∞

$\lim_{n \to \infty} a_n = 2 + \lim_{n \to \infty} \dfrac{(-1)^n}{n} = 2 + 0 = 2.$ The correct answer is D.

---

107. If f is a continuous function on $[a, b]$ and

$f(x) > 0$ for all x in $[a, b]$, then $\int_a^b f(x)dx$

represents the

(A) length of the curve $y = f(x)$ between $x = a$ and $x = b$
(B) area of the region bounded by the graphs of $y = f(x)$, $x = a$, $x = b$, and the x-axis
(C) area of the region bounded by the graphs of $y = f(x)$, $y = a$, $y = b$, and the y-axis
(D) volume of the solid generated by revolving about the x-axis the region bounded by the graphs of $y = f(x)$, $x = a$, $x = b$, and the x-axis
(E) volume of the solid generated by revolving about the y-axis the region bounded by the graphs of $y = f(x)$, $x = a$, $x = b$, and the x-axis

Since $f(x) > 0$ for all x in $[a, b]$, the integral represents the area of the region bounded by the graphs of $y = f(x)$, $x = a$, $x = b$, and the x-axis. The correct answer is B.

---

108. The graph of the equation
$x^2 + 4y^2 + 4x - 24y + 36 = 0$ is

(A) a circle
(B) a parabola
(C) an ellipse
(D) a hyperbola
(E) a point

The equation $x^2 + 4y^2 + 4x - 24y + 36 = 0$ is equivalent to

$$(x^2 + 4x + 4) + 4(y^2 - 6y + 9) = 4$$

$$\frac{(x + 2)^2}{4} + \frac{(y - 3)^2}{1} = 1.$$

This is the standard form of the equation of an ellipse with center $(-2, 3)$, a major axis of length 4, and a minor y-axis of length 2. The best answer is C.

109. The automobile license numbers assigned in a certain county consist of 2 letters followed by 3 digits, and any letter or digit may be repeated. If the letters I and O are excluded, and the first digit cannot be zero, how many different license numbers can be assigned?

(A) $24 \cdot 24 \cdot 9 \cdot 10 \cdot 10$
(B) $26 \cdot 25 \cdot 9 \cdot 9 \cdot 9$
(C) $26 \cdot 25 \cdot 10 \cdot 9 \cdot 8$
(D) $24 \cdot 24 \cdot 9 \cdot 9 \cdot 9$
(E) $24 \cdot 23 \cdot 9 \cdot 8 \cdot 7$

Since I and O are excluded from the letters that can be used, there are 24 possibilities for each of the two letters that can be used. The only restriction on the digits is that 0 cannot be used as the first digit. Therefore, there are 9 possibilities for the first digit and 10 possibilities for each of the remaining two digits. Since the selection of each letter or digit is independent, it follows that it is possible to assign $24 \cdot 24 \cdot 9 \cdot 10 \cdot 10$ different license numbers. The correct answer is A.

---

110. If f and g are inverse functions, which of the following is NOT necessarily true?

(A) $f(x) \cdot g(x) = 1$
(B) $f(g(x)) = x$
(C) $f(g(x)) = g(f(x))$
(D) Both f and g are one-to-one functions.
(E) The domain of g is the range of f.

Two functions f and g are said to be inverse functions if the composition of the two functions $f \circ g$ and $g \circ f$ are each the identity function. Thus, $f(g(x)) = x$ and $g(f(x)) = x$, and both B and C are true. To show that f is one-to-one, it is necessary to show that $f(x_1) = f(x_2)$ implies $x_1 = x_2$. Since $f(x_1) = f(x_2)$, it follows that $g(f(x_1)) = g(f(x_2))$. But $g(f(x_1)) = x_1$ and $g(f(x_2)) = x_2$. Thus $x_1 = x_2$ and f is one-to-one. In a similar manner, it can be shown that g is one-to-one and D is true. Choice A is not necessarily true. For example, if $f(x) = x + 1$, then $g(x) = x - 1$, since $f(x - 1) = g(x + 1) = x$. So f and g are inverse functions, but $f(x)g(x) = (x + 1)(x - 1) = x^2 - 1$, which is not equal to 1 unless $x = \pm\sqrt{2}$. Therefore, the best answer is A.

111. If $m < 0$ and $n > 0$, which of the following could be the graph of $y = mx^3 + n$?

(A)

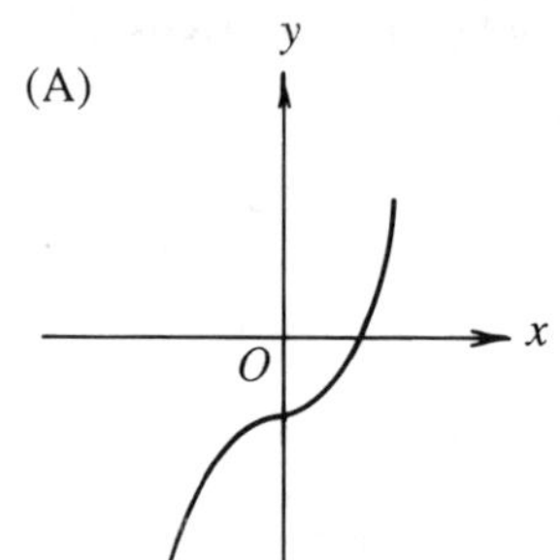

(B)

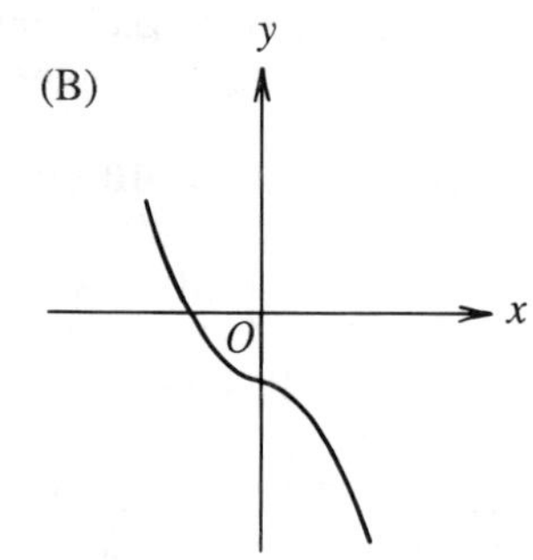

(C)

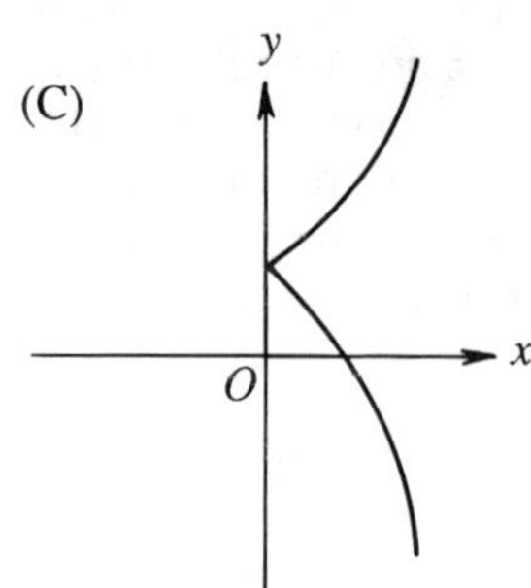

(D)

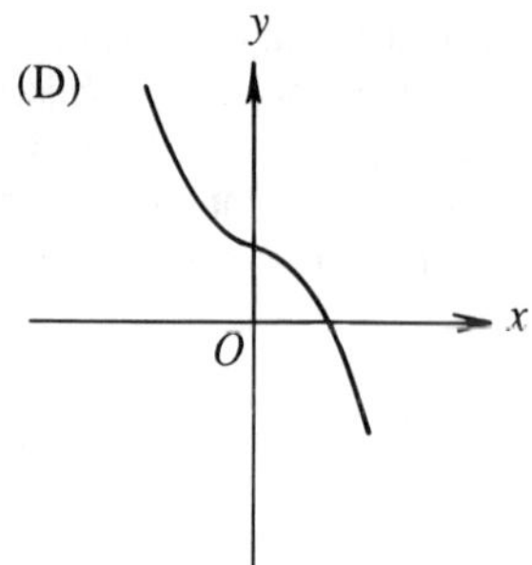

(E)

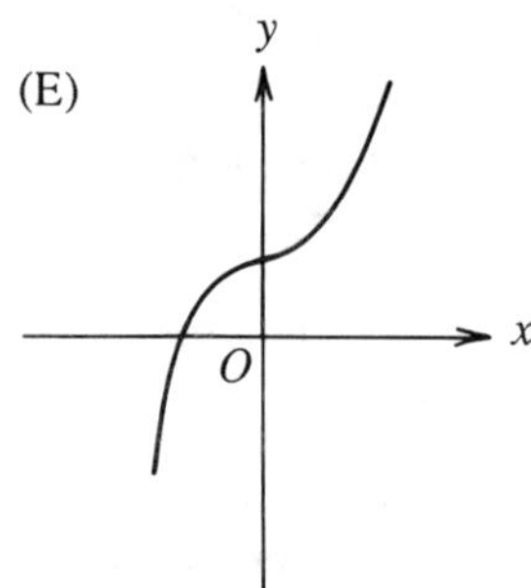

If $y = mx^3 + n$, then $y' = 3mx^2$ is the equation for the slope of the curve, and since $m < 0$, the slope is always less than or equal to 0. Since only choices B and D have negative slope, one of these must be the answer. Since $n > 0$, the y-intercept must be positive. Therefore, the best answer is D.

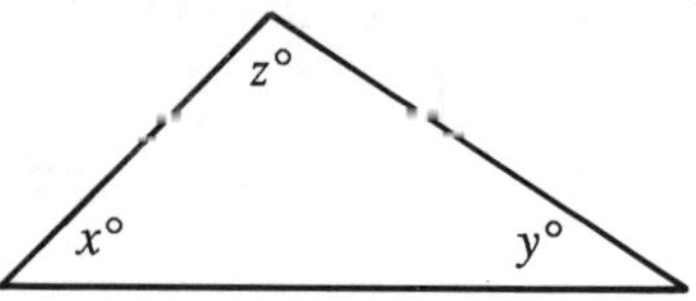

Note: Figure not drawn to scale.

112. In the triangle above, if $60 < y < 110$, then

(A) $60 < x < 110$
(B) $70 < x < 110$
(C) $60 < x - z < 110$
(D) $70 < x + z < 120$
(E) $70 < x - z < 120$

Since $x + y + z = 180$, if $y > 60$, then $x + z < 180 - 60 = 120$. If $y < 110$, then $x + z > 180 - 110 = 70$. Thus, $70 < x + z < 120$, and the best answer is D. A counterexample showing that the other inequalities do not necessarily hold is the case where $y = 61$, $x = 1$, and $z = 118$.

___

113. The amount to which $1,000 will grow in 5 years at a 6 percent annual interest rate compounded semiannually is given by

(A) $1,000(1 + 0.03)^{10}$
(B) $1,000(1 + 0.03)^5$
(C) $1,000(1 + 0.06)^{10}$
(D) $1,000(1 + 0.06)^5$
(E) $1,000(1 + 0.12)^5$

If the annual interest rate on an investment is 6 percent compounded semiannually, then the amount of the investment will grow by $\frac{1}{2}$ of 6 percent, or 3 percent, at the end of each 6-month period. This is calculated by multiplying the amount at the beginning of the 6-month period of 1.03. Thus, the amount to which the $1,000 will grow at the end of 6 months is 1.03($1,000), and at the end of a year it will grow to $1.03(1.03)(\$1,000) = 1.03^2(\$1,000)$. Following the same reasoning, at the end of r 6-month periods the $1,000 will have grown to $(1.03)^r(\$1,000)$. In 5 years there are ten 6-month periods; therefore the $1,000 will have grown to $(1.03)^{10}(\$1,000)$ in that period. Thus, the correct answer is A.

	Set	Relation
A	Integers	is a multiple of
B	Real numbers	is greater than or equal to
C	All sets of integers	is a subset of
D	Triangles	is similar to
E	Lines	is perpendicular to

114. Which of the five relations listed above represents an equivalence relation?

(A) *A*
(B) *B*
(C) *C*
(D) *D*
(E) *E*

A relation R on a set is an equivalence relation if it is reflexive, symmetric, and transitive. Since none of the relations *A*, *B*, and *C* is symmetric (that is, $a\,R\,b$ does not imply $b\,R\,a$), none of these is an equivalence relation. With respect to relation *D*, a triangle is similar to itself (reflexive property); triangle *A* is similar to triangle *B* implies that triangle *B* is similar to triangle *A* (symmetric property); and triangle *A* similar to triangle *B* and triangle *B* similar to triangle *C* imply that triangle *A* is similar to triangle *C* (transitive property). Thus, *D* is an equivalence relation. With respect to relation *E*, neither the reflexive nor the transitive property holds. Therefore, the correct answer is D.

---

115. From a point 1,000 meters from the base of an airport tower, the angle of elevation to the top of the tower is 15 degrees. If h is the height, in meters, of the tower, then $h =$

(A) $1{,}000 \tan 15°$

(B) $\dfrac{1{,}000}{\tan 15°}$

(C) $1{,}000 \sin 15°$

(D) $\dfrac{1{,}000}{\sin 15°}$

(E) $1{,}000 \cos 15°$

The following figure represents the information given in the problem.

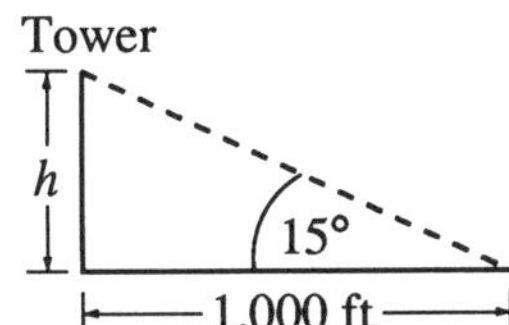

Now $\tan 15° = \dfrac{h}{1{,}000}$ and $h = 1{,}000 \tan 15°$. The correct answer is A.

116. Recommendations made by the National Council of Teachers of Mathematics in their publication *Curriculum and Evaluation Standards for School Mathematics* include which of the following?

   I. The mathematics curriculum should emphasize problem-solving.
   II. All students should have access to calculators throughout their school mathematics program.
   III. There should be increased emphasis on drill with numbers apart from problem contexts.

(A) I only
(B) II only
(C) III only
(D) I and II
(E) II and III

Pages 5 and 6 of the *Curriculum and Evaluation Standards for School Mathematics* state the "New Goals for Students." The third goal listed is "Becoming a mathematical problem-solver" and states: "The development of each student's ability to solve problems is essential if he or she is to become a productive citizen. We strongly endorse the first recommendation of *An Agenda for Action* (NCTM 1980): 'Problem-solving must be the focus of school mathematics, . . . etc.' " The importance of building the curriculum around problem-solving activities is stressed throughout the *Standards*. On page 8 of the *Standards*, the authors stated their belief that "appropriate calculators should be available to all students at all times." Thus, both I and II are recommendations of the NCTM; however, III, "the increased emphasis on drill with numbers apart from problem contexts," was not a recommendation and is, in fact, antithetical to the spirit of the *Standards*. The best answer is D.

---

117. What is the fifth term in the binomial expansion of $(x + 1)^8$ when the powers of x are written in descending order?

(A) $3{,}024x^4$
(B) $70x^4$
(C) $56x^5$
(D) $35x^4$
(E) $21x^5$

If $(a + b)^n$ is expanded by the rules of the binomial theorem, the coefficient of the term containing $a^{n-r}b^r$ is given by $\dfrac{n!}{r!(n-r)!}$. Therefore, the fifth term of the expansion of $(x + 1)^8$, written in descending powers of x, is the term that contains x^4 and has coefficient $\dfrac{8!}{4!4!} = 70$. Thus, the required term is $70x^4$ and the correct answer is B.

118. A pack of cards is numbered from 1 to 36 inclusive. If one card is selected at random, what is the probability that the number of the card is a multiple of 4 or 6 but not a multiple of 12 ?

(A) $\dfrac{7}{12}$

(B) $\dfrac{1}{2}$

(C) $\dfrac{1}{3}$

(D) $\dfrac{1}{4}$

(E) $\dfrac{1}{6}$

Of the numbers from 1 to 36, inclusive, 9 are multiples of 4 and 6 are multiples of 6; but the numbers 12, 24, and 36 belong to both sets and are not to be included in either set. Therefore, the number of possible successes is $(9-3)+(6-3)=9$, and the probability of obtaining one of these successes is $\dfrac{9}{36}=\dfrac{1}{4}$. The correct answer is D.

---

119. The scores of 500 students on an examination were normally distributed with a mean of 70 and a standard deviation of 10. Letter grades were assigned as follows:

        Above 86 - A
        81-86 - B
        60-80 - C
        53-59 - D
        Below 53 - F

Approximately how many students received a grade higher than C ?

(A)   50
(B)   80
(C)   100
(D)   120
(E)   160

The figure below is a graph of a normal distribution with mean 0 and standard deviation σ. Note that the vertical line at the mean (and the median) splits the distribution in half. Furthermore, 34 percent of the cases (letter grades) fall between the mean (70, in this problem) and one standard deviation above the mean $(70+10=80)$. The number of students who received a grade higher than C is precisely the number who scored more than one standard deviation above the mean (that is, more than 80). This proportion of the scores is $50\% - 34\% = 16\%$, and 16% of 500 students is 80. The best answer is B.

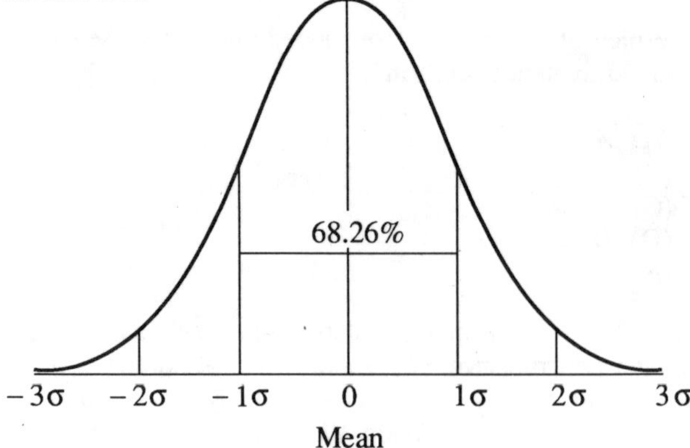

Standard Normal Distribution

---

120. If $y = x^2$ and $x^y \cdot y^x = x^w$ for $x > 0$, then, in terms of x, $w =$

(A) $x^2 + 2x$
(B) $x^2 + x + 2$
(C) $x^2 + 2^x$
(D) $2x^2$
(E) $2x^3$

Since $y = x^2$, one can substitute x^2 for y in the equation

$x^y y^x = x^w$

$x^{x^2}(x^2)^x = x^w$

$x^{x^2+2x} = x^w$. Since $x^{(x^2+2x)} = x^w$, it follows that $x^2 + 2x = w$.

The correct answer is A.